1,000,000 Books

are available to read at

www.ForgottenBooks.com

Read online
Download PDF
Purchase in print

ISBN 978-1-330-24893-5
PIBN 10000621

This book is a reproduction of an important historical work. Forgotten Books uses
state-of-the-art technology to digitally reconstruct the work, preserving the original format
whilst repairing imperfections present in the aged copy. In rare cases, an imperfection in
the original, such as a blemish or missing page, may be replicated in our edition. We do,
however, repair the vast majority of imperfections successfully; any imperfections that
remain are intentionally left to preserve the state of such historical works.

Forgotten Books is a registered trademark of FB &c Ltd.
Copyright © 2018 FB &c Ltd.
FB &c Ltd, Dalton House, 60 Windsor Avenue, London, SW19 2RR.
Company number 08720141. Registered in England and Wales.

For support please visit www.forgottenbooks.com

1 MONTH OF FREE READING

at

www.ForgottenBooks.com

By purchasing this book you are eligible for one month membership to ForgottenBooks.com, giving you unlimited access to our entire collection of over 1,000,000 titles via our web site and mobile apps.

To claim your free month visit:
www.forgottenbooks.com/free621

* Offer is valid for 45 days from date of purchase. Terms and conditions apply.

English
Français
Deutsche
Italiano
Español
Português

www.forgottenbooks.com

Mythology Photography **Fiction** Fishing Christianity **Art** Cooking Essays Buddhism Freemasonry Medicine **Biology** Music **Ancient Egypt** Evolution Carpentry Physics Dance Geology **Mathematics** Fitness Shakespeare **Folklore** Yoga Marketing **Confidence** Immortality Biographies Poetry **Psychology** Witchcraft Electronics Chemistry History **Law** Accounting **Philosophy** Anthropology Alchemy Drama Quantum Mechanics Atheism Sexual Health **Ancient History Entrepreneurship** Languages Sport Paleontology Needlework Islam **Metaphysics** Investment Archaeology Parenting Statistics Criminology **Motivational**

The Blacksmith's Guide

VALUABLE INSTRUCTIONS ON FORGING,
WELDING, HARDENING, TEMPERING,
CASEHARDENING, ANNEALING,
COLORING, BRAZING, AND
GENERAL BLACKSMITHING

By J. F. SALLOWS

FIRST EDITION

THE TECHNICAL PRESS
BRATTLEBORO, VT.
1907

COPYRIGHT, 1907
BY
J. F. SALLOWS

E. L. HILDRETH & CO.
PRINTERS
BRATTLEBORO, VT.

DEDICATION

TO MY YOUNG SON
FRANK HAMILTON SALLOWS
FOURTH IN DESCENT
IN A FAMILY OF BLACKSMITHS
MAY HE BECOME FIRST
IN MANLINESS AND SKILL

INTRODUCTION.

In offering this book to my fellow craftsmen, I do not wish it to be inferred that I consider myself the only one who knows how to do the work described in its pages. In 27 years' experience at blacksmithing, however, working in nearly all kinds of shops, including horseshoeing, marine, railroad, printing press, sawmill machinery and automobile shops, I have had opportunities seldom obtained by the average smith. I therefore hope that the book will not only help the young men in the trade, but some of the older blacksmiths as well; and since much attention has been given to the subjects of hardening, tempering, casehardening, coloring, etc., I believe it will also prove useful to machinists and toolmakers. Part of the matter upon these latter subjects was contributed, in somewhat different form, to the columns of *Machinery*, but most of the material is here used for the first time.

Everything in these pages is from actual experience and I am ready at all times to answer any question on any subject that is not fully understood by the reader; but I have, tried to make everything so plain that the average blacksmith can readily understand the methods explained.

The man who gives the best satisfaction is the one to get the highest wages and I am confident that one who follows the directions in this book *will* give satisfaction. The methods described show how to become a rapid and an independent workman, which is the kind employers are looking for, although this kind seems hard to find at the present time, especially among the younger blacksmiths.

Fourteen years of my blacksmithing experience have been spent as foreman and during this time I have observed that blacksmiths in general have but a small chance to learn anything more than they can dig up in their own daily toil. A young man from the farm can go into a machine shop, start in by running a drill press, then a lathe, and by reading and strict attention to business he will soon become a fairly good machinist. It is not so with a blacksmith, and especially a machine blacksmith, who usually has difficulty in acquiring full knowledge of his trade. Something should be done to assist the young men who are willing to learn the trade most difficult to learn—that of blacksmithing.

"For since the birth of time, throughout all ages and nations,
Has the craft of the smith been held in repute by the people."

[*Longfellow.*]

J. F. SALLOWS.

Lansing, Mich., July, 1907.

CONTENTS.

CHAPTER I. MACHINE FORGING.......................... 1
Reading Drawings—Instructions Should be Clear—Arranging Forges in a Shop—The Anvil Block—Cutting Off Steel—Heating and Forging—Tongs—Heading Tool—Bending Fork—Bevel Set—Bolts for Planer—Key Puller—Open End Wrench—Socket Wrench—Spanner Wrench—Turnbuckle or Swivel—Crankshaft—Making a Square Corner in Heavy Stock—Making a Double Angle—Directions for Welding—Scarf Weld—Two Heats to Make a Weld—Butt Weld—Lap Weld—Cleft Weld—Jump Welding—Welding a Ring—Welding Solid Ends in Pipe—The Use of "Dutchmen."

CHAPTER II. TOOL FORGING............................. 40
Systematic Arrangement of Work—Care in Heating—Patterns—Bevel Set—Cold Chisels—Cape Chisels, Grooving Chisels, Etc.—Screw Driver—Tools from Files—Cutting-off Tool—Side Tools—Boring Tool—Threading Tools — Roughing Tool — Finishing Tools — Centering Tools—Diamond Point—Round Nose Tool—Brass Turning Tool—Rock Drill—To Make Tool Steel Rings or Dies Without Welding.

CHAPTER III. HARDENING AND TEMPERING................ 66
Proper and Improper Heating—Drawing the Temper—Charcoal Fire—Home-made Oven—Color Charts—Hardening Cold Chisels—Center Punches—Lathe and Planer Tools—Hardening Milling Cutters—Hardening Formed Cutters—Hardening a Thin Cutter—The Treatment of Reamers—Hardening Taps—Treatment of Punches and Dies—Threading Die—Treatment of a Broach—Shear Blades—Hardening Large Rolls—Tempering Springs—Drill Jig Bushings—Tempering a Hammer Head—Tempering Fine Steel Points—Shrinkage and Expansion—Annealing—General Directions for Hardening.

CHAPTER IV. HIGH-SPEED STEEL..........................107
To Distinguish High-Speed Steel—Cutting Off High-Speed Steel—Treating Self-Hardening Steel—Forging Air-Hardening Steel—Device for Air Hardening—Heating for Hardening—Hardening with Cyanide—Hardening Milling Cutters and Other Expensive Tools—Hardening Long Blades—Annealing High-Speed Steel.

CHAPTER V. CASEHARDENING AND COLORING...............118
The Furnace—Packing for Casehardening—Directions for Casehardening—Tank for Casehardening Work—Tools from Machine Steel—Pack Hardening—Pack Hardening Long Pieces—Testing Work—Pack Hardening Thin Cutters—Cyanide Hardening—Mottling and Coloring—How to Get the Charred Bone and Leather—Coloring with Cyanide—Coloring by Heat Alone.

CHAPTER VI. BRAZING—GENERAL BLACKSMITHING.......134
Brazing — Brazing Furnace — Spelter — Directions for Brazing—Brazing Cast Iron—Brazing a Small Band Saw—Bending Gas Pipe—To Straighten Thin Sheet Steel—The General Blacksmith and Horse Shoer—Repairing Plowshares—Shoeing to Prevent Interfering—Horseshoe Vise—Shoeing for Contracted Feet—Advice to Foremen.

APPENDIX.
Table of Decimal Equivalents—Reproductions in Colors Showing Coloring of a Hardened Wrench—Colored Heat Chart—Colored Temper Chart—Working Drawings of a Casehardening Furnace.

THE BLACKSMITH'S GUIDE.

CHAPTER I.

MACHINE FORGING.

Reading **Drawings.**—Machine forging is the simplest class of blacksmithing and if the smith understands drawings he will not find this kind of work difficult; but if he does not understand drawings he may better devote his time to some other class of smithing. The main point to bear in mind is to leave

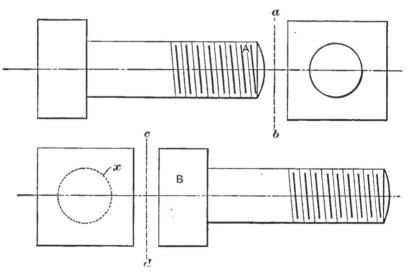

Fig. 1. Arrangement of Views in Working Drawings.

stock for finishing wherever called for. Drawings are always marked with the letter *f* or otherwise where required to be finished; or if a piece is to be machined all over, as in Fig. 53, it will be so stated on the drawing.

By using a common square head bolt, Fig. 1, I will give an illustration which I think will help the blacksmith in reading mechanical drawings. The upper half of the sketch shows side and end views of the bolt, with the end view at the right; while in the lower half the end view is at the left. In each case the end view represents the bolt as it would appear if looked at from the end at which this view is placed. Thus, in the upper part of the sketch the end view shows the bolt as it would appear if looked at from the right-hand end and in the lower part as it would appear from the left-hand or head end.

This will be better understood by supposing the sheet to be folded on the dotted line $a\,b$ in the upper half of the sketch, which will bring the end view into the correct position to represent the end A of the bolt as it would appear if looking directly at that end. Again, by folding the sheet on the dotted line $c\,d$ in the lower half of the sketch, the end view will be in the correct position to represent the other end B of the bolt; but in this view the head would be the only part visible and would completely cover up the shank, so the circle x representing the shank should here be drawn dotted instead of solid, as in the upper view.

Throughout this book the illustrations are in the form of working drawings so far as practicable, showing side, end or top views of the pieces, as necessary. In each case these different views are correctly placed in relation to each other; and by a study of these it will be clear why certain lines are dotted in some views and solid in others, as in the two end views of the bolt just used as an illustration.

Another thing I want to call to the attention of blacksmiths is the dimension of a radius as given on a drawing. **Radius** means one half the diameter of a circle; but time and again I have seen smiths, when getting a job of work such as Fig. 2, where the size of a circle is given by "¾-inch **R**," form the part over a ¾-inch rod instead of using a 1½-inch rod as they should. Many smiths do not understand this, but it is important, and by observing this it will save the humiliation of having a piece of work returned to be operated on the second time.

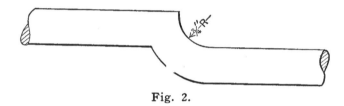

Fig. 2.

At the end of the book will be found a table of fractions with their decimal equivalents which will be useful in working from drawings. Oftentimes a drawing comes to the smith with the dimensions given in decimals, thus: "2.3125 inch," which is the same as 2⁵⁄₁₆ inches.

Instructions Should be Clear.—It makes a great saving in the time a smith will devote to a job if he is informed what it is for. Sometimes he will take great pains with work that would answer every purpose if made with considerably less care, as would be the case with a repair job. I have known foremen to request a smith to take as much pains with a forging of this character as with one that it was necessary to have finished carefully, with the result that twice as much time was put upon the job as would have been required if the smith had known what it was for.

For example, in Fig. 3 are two views of a piece bent at a right angle. A drawing will come from some department looking like No. 2, when if made as in No. 1 it would be just as good and certainly would be stronger. Of course there are cases where a piece of

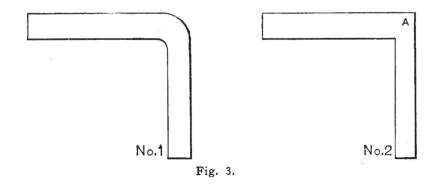

Fig. 3.

work must be square on the outside corner *A*, Fig. 3; but all blacksmiths know the difference in the time consumed in making the two pieces and for lack of forethought on the part of the foreman smith there is a vast amount of time wasted in just such cases.

Arranging Forges in a Shop.—It is sometimes the custom to arrange forges in a smith shop in a straight line along one side near the wall, giving little thought to the inconvenience caused the workman by placing the forges in a haphazard way; and the financial loss that results from locating the smiths where they work at a disadvantage. Fig. 4 shows a simple and at the same time a convenient way to arrange the forges in a smith shop. They are placed in a semi-circle, which gives all the smiths the same chance at the steam and trip hammers, located at *A* and *BB* respectively. If only one or two smiths in a shop are allowed to use these hammers, the rest feel slighted and have reason to feel so, as there is no smith in a large shop who does not have occasion to use the hammers at one time or another. By having the fires so arranged, it does away with the "dog in the manger" feeling usually displayed by the smith whose fire is nearest the hammer.

The Anvil Block.—It is the custom in a great many shops to fasten the anvil to a large wooden block set in a big hole dug in the ground. This type of foundation has the objection that if you want to move the

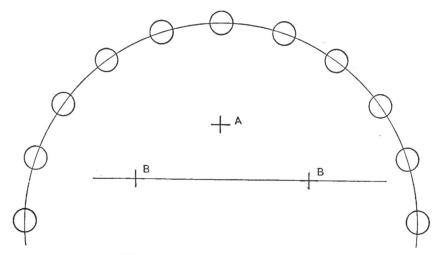

Fig. 4. Arrangement of Forges.

anvil at any time and perhaps later return it to its old location it will be found quite a task to do so. Fig. 5 shows top and side view of a hollow cast-iron anvil base that will be found more satisfactory in all respects. The sides taper from C to D, as indicated, and there is a wide flange B around the bottom. The metal is about one inch thick and there are large round holes in the sides and top.

6   THE BLACKSMITH'S GUIDE

Fig. 6 shows the anvil mounted on the base. A hard wood plank *A* is put between the base and the anvil and by having four thin steel straps *B*, one on each side of the anvil bottom, and four ⅝-inch hook

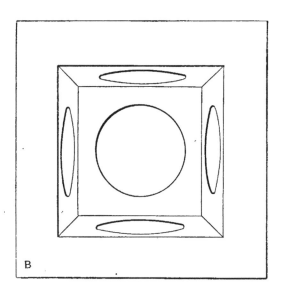

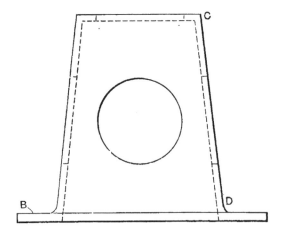

Fig. 5. Anvil Base.

bolts *C*, the anvil can be securely bolted to the base and the anvil, together with its base, can be readily moved to any part of the shop desired, since the base rests directly on the shop floor without any other foundation.

Cutting Off Steel.—Cutting-off tools for cutting off round steel on the anvil in the blacksmith shop are shown in Fig. 7. The bottom cutter is at *A*, the upper

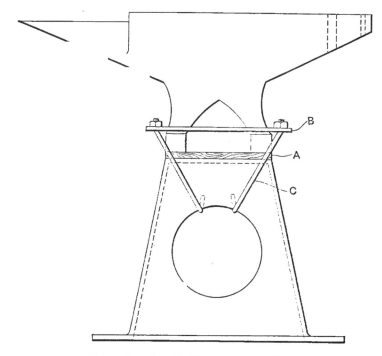

Fig. 6. Anvil Mounted on Base.

or hand cutter at *B*, and at *C* is a round bar in position to be cut off. The tools should be tempered the same as a cold chisel, as described later, and they will be found a very useful addition to the blacksmith's outfit.

To break large bars of steel under the steam hammer, after nicking around the bar as shown at *X*, Fig. 8, place two small, round pieces of steel *AA* on the bottom die of the steam hammer; then rest the bar which is to be broken on these pieces, and place another small piece of steel on top of the bar in the center, at *B*. By striking a good solid blow the bar will be easily broken. A bar will break more readily if a little water is first poured around the nicks.

Heating and Forging.—Carelessness in heating and hammering either machine or tool steel is the cause of many forgings proving defective after being machined. If a piece of round steel is heated too quickly

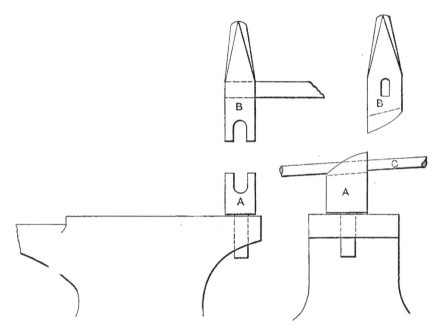

Fig. 7. Cutting-off Tools.

and drawn down under the hammer it will be concave as at *A*, Fig. 9. This causes a parting of the metal, and when machined the piece will show checks and cracks of considerable size. The same results will follow if not properly hammered, even if the piece is carefully heated. If the blows are not heavy enough to affect the metal at the center of the bar the outside will draw

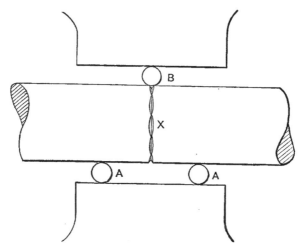

Fig. 8. Breaking Large Stock.

out, leaving the center as at *A;* but if heated uniformly and drawn down under a hammer that gives blows heavy enough to affect the center of the bar, the end will be convex as at *B*, making a much stronger job. In the days when wrought iron was used for machine forging we could weld up such defects as the foregoing and they would not be noticed; but with steel it is different and the defects cannot be hidden by weld-

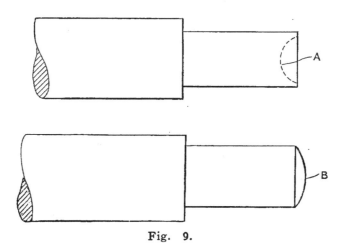

Fig. 9.

ing. There is so much trouble from bad judgment in heating and working steel that a tool machined from the bar and hardened and tempered will give better satisfaction than if forged by the average smith. This caution about heating and hammering applies to work done on the anvil with the hand hammer as well as under a steam or trip hammer. The steel should be properly heated and the blows should be rapid and heavy enough to affect the center of the bar, even when making so small a thing as a chisel or a center punch.

Some blacksmiths depend too much on the file for finishing a forging. Now, in a shop where there are emery wheels, shapers and planers, lathes and milling machines, there is no excuse for a blacksmith spending half his time filing his work. Of course, in cross roads blacksmith shops this cannot be avoided; but I am now referring to shops connected with the large factories. I like to see a smith throw his work down on the floor, hot, right from the hammer, and not try to rub all the stock off when it has to be filed or machined to a finish in some other department.

Tongs, like fullers, flatters and swages, are made by drop forging, so that the blacksmith has very little of this kind of work to contend with by hand. In fact, it does not pay to bother to make any but special tongs that are wanted at once and cannot be purchased or that it would take some time to get by sending for them. A quick and handy way to make a pair of light tongs is indicated in Fig. 10. Take a piece of flat stock of the size required and flatten the ends as shown in the two views, A and B. Then punch holes $x\ x$ and cut along the dotted line in lower view, *C*; round up the handles and rivet the two parts together; then heat the jaws to a bright red and cool off, at the same time

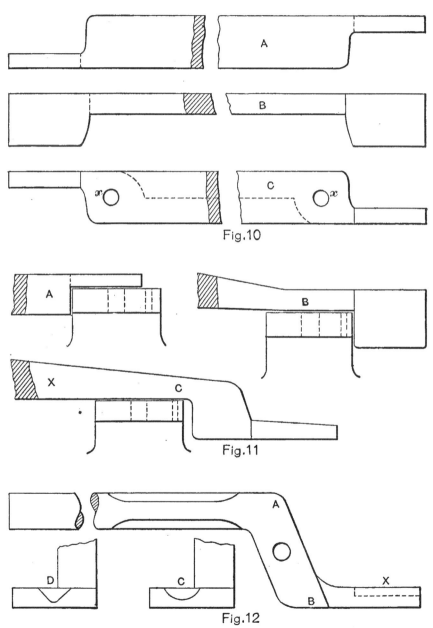

Fig. 10

Fig. 11

Fig. 12

Method of Making Blacksmith's Tongs.

opening and closing until cold. This allows the tongs to work freely instead of binding and being stiff. This should be done with any pair of tongs which gets stiff for want of use.

If larger tongs are to be made, Fig. 11 shows how each jaw can be produced in three operations and in

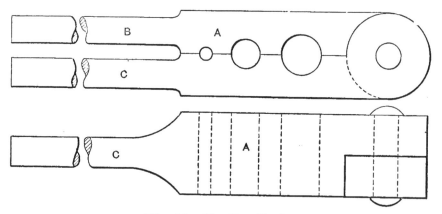

Fig. 13. Heading Tool.

one heat by a smart smith and his helper. The first operation is to flatten down from the square stock to the desired size as at *A*, holding the piece so that the straight side will be toward the square end of the anvil, or to the right hand of the smith, and letting the stock all draw out toward the left end of the anvil. After flattening to the desired thickness turn the arm to the left, giving the piece a quarter turn to the position shown at *B*, and flatten the stock where the hole is to be punched for the rivet. Now turn the arm holding the work to the left again and shove the jaw far enough over the anvil so that the piece can be forged in position *C*. Finally change ends and draw out the part marked *X*, and weld on the handle.

The size of a pair of tongs is governed by the length of the part from *A* to *B* in Fig. 12; and also remember that all flat jaw tongs should be grooved with a V or oval as in the end views *C* and *D* in Fig. 12. This allows the smith to hold any shape stock that comes his way. The indentation should be only part way along the jaw as indicated by the dotted line at *X*.

Heading Tool.—Fig. 13 shows a handy heading tool for special work that cannot be done in the common heading tool. The heavy part *A* should be of tool steel and the handles *B* and *C*, which are of machine steel, welded on. The tool can be made in different sizes and different sized holes drilled when the heading tool is closed, to accommodate work of different dimensions. The two parts of the tool should be fastened together with a strong rivet.

Bending Fork.—In Fig. 14 is a side and edge view of a smith's bending fork, a very handy tool and one easily made. It has a square shank *B* to fit in the anvil and two round prongs *A A* by which the bending is accomplished. A flat bar *C* is shown between the prongs in the right position for bending.

In Fig. 15 is a bending fork to be used in the hand the same as a wrench. This tool is used a great deal by machinists and assemblers. In making this fork weld a round piece *A* to round piece *B*, Fig. 16, by a cleft weld and then turn up the end as indicated by the dotted line. A stronger fork can be made without a weld as shown in Fig. 17. Take a flat bar of a size suited to the size of tool wanted and draw out the round handle *A;* and then cut out piece *B* along the dotted lines. This is perhaps the best way to make a bending fork if a power hammer is at hand.

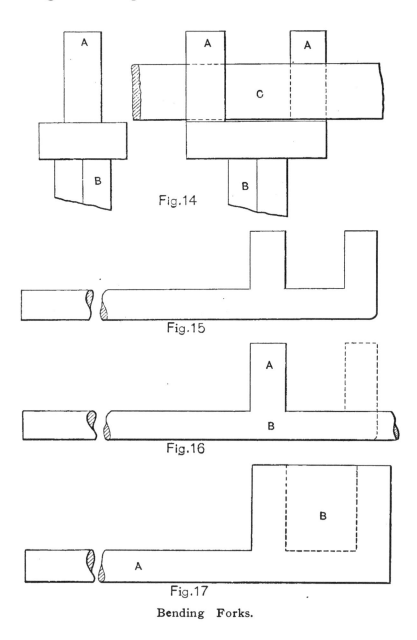

Bending Forks.

Bevel Set.—In Fig. 18 is a bevel bottom set that takes the place of a fuller and must be used in **conjunc-**

MACHINE FORGING

tion with the top bevel set in Fig. 19. These sets are better than either fullers or chisels for the class of

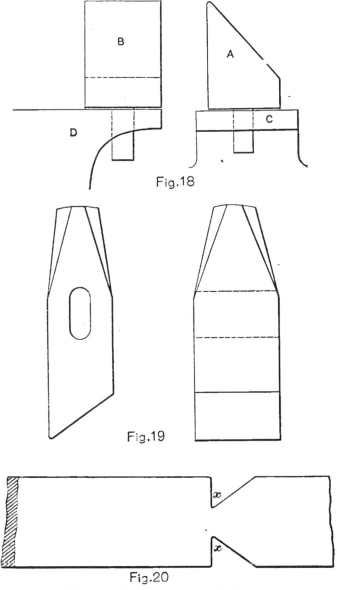

Top and Bottom Bevel Sets.

work that is started as shown in Fig. 20, where the part marked *A* is to be drawn out. It makes a neat small fillet *x* and produces a straight, square shoulder on the work. In Fig. 18 *A* is the end view and *B* the

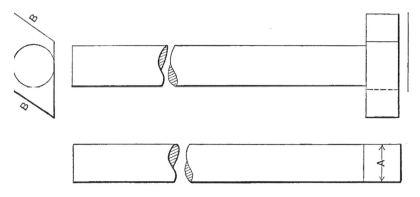

Fig. 21. Planer Bolt.

side view of the bottom set, while *C* is the end view and *D* the side view of the anvil.

Bolts for Planers.—Fig. 21 shows a bolt for dropping into the slots in any part of a planer bed, thus doing away with starting at the end of the planer to

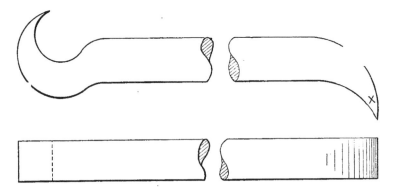

Fig. 22. Key Puller.

slide the bolt to place, as is done with ordinary square head bolts. When forging this bolt, make a T-head of the proper width at *A* to fit the planer slot; then cut off the corners *B B*. This bolt can be dropped into place and turned to the right so that faces *B B* will tighten against the inside walls of the planer slots as the nut is screwed up. If quite a number of these bolts of different lengths are furnished for each planer in the shop it will lessen the cost of getting out the planer work and will make the planer hand happy.

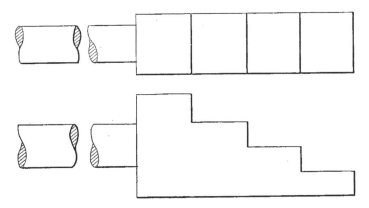

Fig. 23. Assistant for Key Puller.

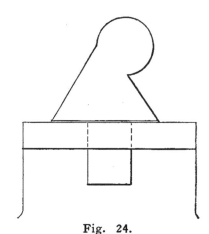

Fig. 24.

Key Puller.—A useful tool for machine shop, tool room, erecting room, repair department and shipping room is the jimmie-bar or key puller and its assistant. Fig. 22 shows side and front view of jimmie-bar and Fig. 23 side and front view of the assistant. Fig. 24

shows the bottom fuller in anvil, used in making the jimmie-bar. The bar is made in different lengths and sizes, from ⅜ inch octagon to 1½ inch round, and

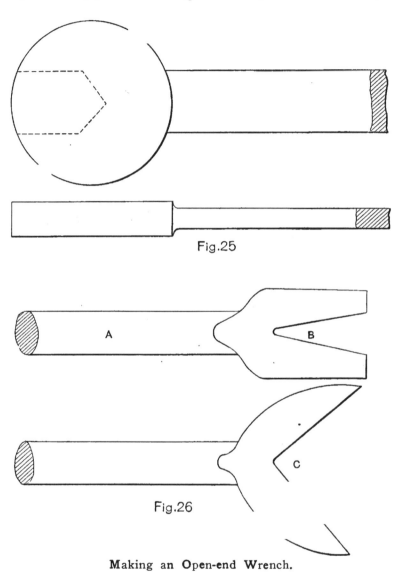

Fig.25

Fig.26

Making an Open-end Wrench.

from 12 inches to 36 inches in length. When pulling a key, first start it with the end of bar marked *X;* then use the other end of bar. If the key cannot be removed with the bar alone, place the assistant between the hub and key and you can then remove the key easily. This is much better than looking for half an hour for a block to place between the hub and key, as is usually done.

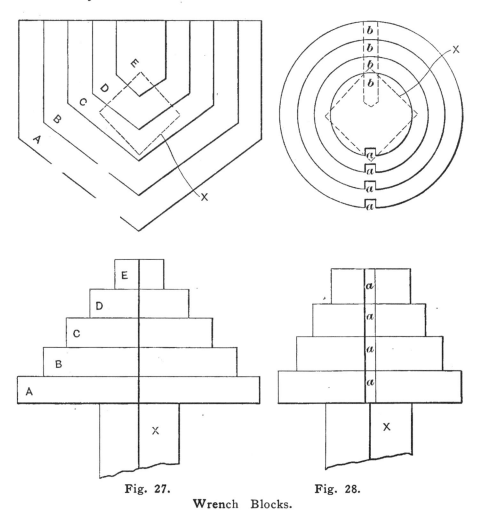

Fig. 27. Fig. 28.
Wrench Blocks.

Open End Wrench.—In making an open end or S wrench, it is the usual practice to draw out the handle and round up the end as in Fig. 25, and then cut out the opening to the size wanted as indicated by the dotted lines. This requires a large piece of stock, and while it may be best to follow this method in some cases, it is not always necessary. A convenient way is to take a small piece of steel, draw out the handle *A,* Fig. 26, split open the end as at *B,* spread out to the right angle as at *C,* and then bend the jaws of the

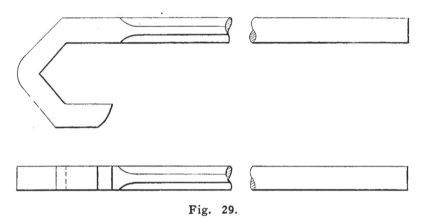

Fig. 29.

wrench to the required size and shape. This may be most conveniently done by having a wrench block as shown in Fig. 27, which can be made either of cast iron or steel and arranged to fit in the square hole in the anvil or in a large swage block. By having several steps or shoulders the wrench block will answer for all sizes of wrenches used in the average shop. The one in Fig. 27 is in five different sizes, from *A* to *E,* and is one of the handiest tools to be found in a smith shop. The square shank designed to fit the hole in the anvil is at *X* in Fig. 27. In Fig. 28 is a wrench block for a spanner wrench, which will be described shortly.

A very strong and handy wrench can be made by rounding up part of a square bar and bending as in Fig. 29.

MACHINE FORGING

Socket Wrench.—In making a socket wrench we sometimes draw down a stem on a large piece of steel,

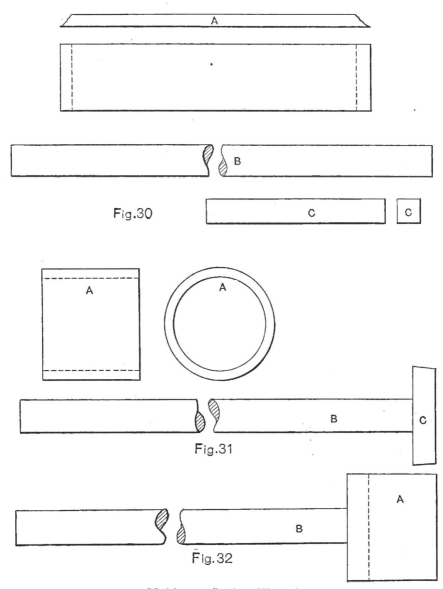

Making a Socket Wrench.

get a hole drilled at the large end and then fit to a mandrel or nut, as the case may be, and weld on a T-handle.

22   THE BLACKSMITH'S GUIDE

The illustrations (Figs. 30-32) show how I make a socket wrench for anything larger than a ½-inch nut. In Fig. 30 assume the piece *A* to be ¼- by 1¼-inch flat stock; *B,* a piece of ⅝-inch round; and *C* a piece

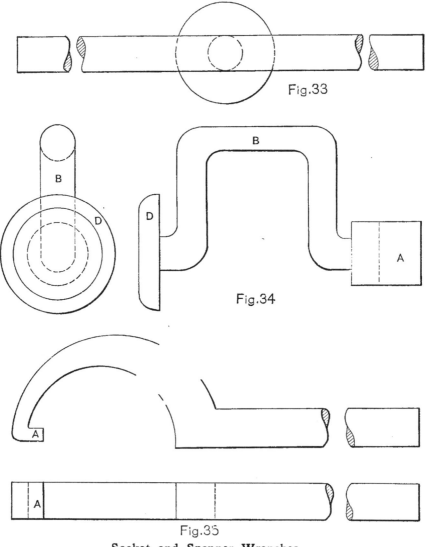

Socket and Spanner Wrenches.

of ⅜-inch square stock. Make a collar from the flat stock as at *A,* Fig. 31; also a collar from the ⅜-inch square stock and weld the latter on the end of the ⅝-inch round bar. (See *C* and *B,* Fig. 31.) Now drive the bar with the collar *C* into the end of the collar *A* and weld and fit to a nut or mandrel, making a piece which looks like Fig. 32. By welding a cross piece on the other end of the ⅝-inch round bar we have a T-handle socket wrench as in Fig. 33. Again, by bending the bar *B* (Fig. 34) and riveting on a wooden block *D* at the end we have a brace which is very handy for putting on nuts rapidly where there is a lot of this work to do.

Spanner Wrench.—Fig. 35 shows a spanner wrench, a tool that blacksmiths are sometimes called upon to make in different sizes. It must be made the proper shape to give satisfaction to the user. In some cases the part marked *A,* Fig. 35, is required to be round, and in others oblong. A block for bending this kind of a wrench was shown in Fig. 28. This form or block (same as in Fig. 27) can be set in the anvil or swage block by having a square shank *X*. Have a row of keyways cut along one side at *a, a,* etc., and a row of different sized holes drilled on the opposite side *b, b,* etc. This forms a handy tool for bending a spanner wrench to the proper size in one heat.

Turnbuckle or Swivel.—Directions will first be given for the swivel, since the same steps are followed in making a turnbuckle. First select a piece of flat stock and bend into a collar like the one in Fig. 36, with the ends separated at *A* by about ⅛ inch, this distance depending somewhat upon the size of swivel that is to be made. This collar is to form the threaded part of the swivel. Cut a groove *y y* on each side of the collar. Now take a piece of round stock of the size required for the yoke, and bend as in Fig. 37, making the distance *x x* between the two ends ⅛ inch less than the distance *y y* between the grooves of the collar in Fig. 36. This is so that when the yoke is sprung over the collar ready for welding it will stay

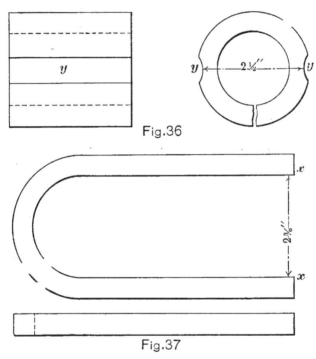

Pieces for Making a Swivel.

in place. Drive the yoke down along the grooves *y y*, take a welding heat and weld as shown in Fig. 38, where *A* is the collar, *B* a mandrel driven into the collar, *C* and *D* swages, and *E* the anvil. By having the mandrel ready to slip in the collar it forms a solid support when welding with the two swages. Fig. 39 shows two views of a completed swivel.

MACHINE FORGING

In making a turnbuckle, first complete a swivel as above directed, and then cut through at *X*, Fig. 38, straighten out the yoke at the ends, and weld to another collar, repeating the operations above outlined for the other end of the turnbuckle.

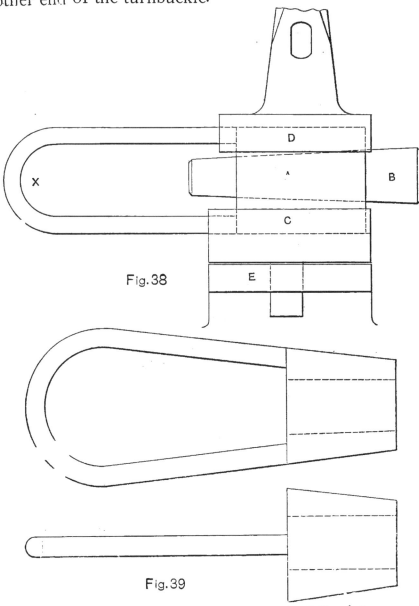

A Completed Swivel and Method of Forging.

Crankshaft.—If called upon to make a crankshaft, as in Fig. 41, with two cranks, first forge with both cranks on the same side, as in Fig. 40; then heat at X and twist into shape. In this way it can be made from smaller stock than as though it were forged at first with the cranks opposite in their correct relative positions.

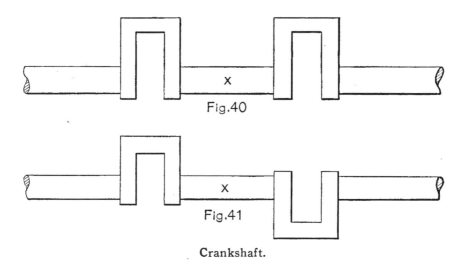

Crankshaft.

Making a Square Corner in Heavy Stock.—I have known blacksmiths to work hard for four or five hours on a job like Fig. 42, made from 2½-inch square stock, when, if they had followed the method here described, it could have been done in half the time and would have given just as good satisfaction. All smiths know how difficult it is to square up the outside corner of a piece made from 2½-inch square stock, as at X, Fig. 42. To make an angle like this, select a piece of square stock of the size called for by the drawing, allowing for finish if required. With the hot chisel cut about half way through the bar as at A, Fig. 43, and bend as shown at *B,* which will open up the outside corner, as indicated at *x,* and give an opportunity to weld in a piece for the purpose of building up a good, solid square corner. To do this, take a bar *C* of

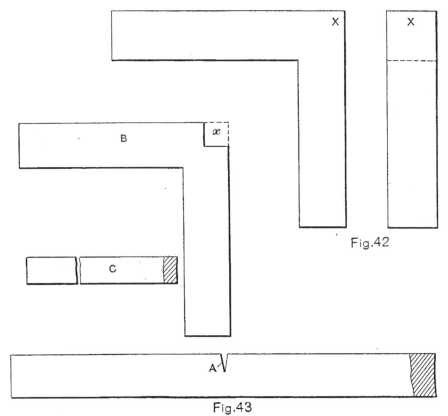

Fig.42

Fig.43

Forging a Square Corner.

smaller square stock, heat and cut one end nearly off. Now even up the sides of the opening in the corner of *B,* take a welding heat on the end of *C,* and weld the piece cut off from the smaller bar into the open corner of *B.* The corner can then be hammered out square and true, as indicated by the dotted lines, and the job will be stronger and done much quicker than by trying to square it by pecking at it until you are tired and get a cold shut in the inside corner of the angle, as you are almost sure to do.

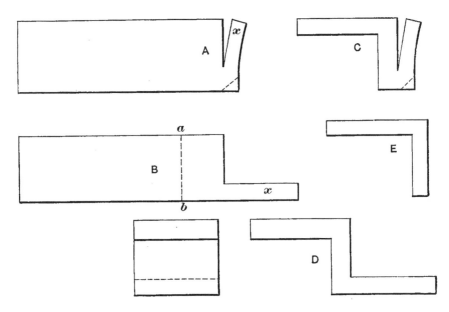

Fig. 44. Making a Double Angle.

Making a Double Angle.—Another troublesome job is to make a double angle from any flat stock over one half inch thick. To do this, take a square bar of the required size and cut as shown at *A*, Fig. 44, and cut off the corner at the dotted line. Then, with a set hammer drive down part *x*, getting the piece into the shape shown at *B*. Finally, cut off the right-hand end, along the dotted line *a b* and repeat the operation for the second angle, as indicated at *C*, thus getting a good, solid forging, like that shown at *D*, where the end view appears at the left and the side view at the right. Even if making a single angle, as at *E*, it is better to make it in this same way if above half an inch thick and two inches wide, for the simple reason that after all your hard work in the usual way of bending and working up a square corner, you will always find a cold shut at the inside corner.

Directions for Welding.—Before going further with machine forgings we will take up the question of welding. I have seen blacksmiths take a dozen heats trying to make a weld and put in six or eight "dutchmen" before they got through with it. The dutchmen can be used to good advantage in some cases, as I will explain further on, but in other cases they should not be used. For good welding the fire must be kept in shape. It should be deep and have a clean, heavy bed underneath the parts about to be welded. Only just enough blast should be used to supply the required heat, and the heating should be done slowly rather than too fast. Underheating will give a bark weld. The pieces will stick on the outside and be open inside, making a deceptive and dangerous piece of work. If the fire is shallow and dirty the oxygen will oxidize the parts, making it impossible to weld them. Fluxes are used to prevent the oxidation of the steel or iron. Any good flux that will exclude the air will answer. About as good as any that I have used, and one of the cheapest, is powder made by pounding up white marble chips. This powder will not fall off and leave the steel exposed to the air. For welding tool steel there is nothing better than borax. In placing pieces in the fire after scarfing and dipping in the flux, always turn them so the scarf will come uppermost, thus keeping the scarf away from the blast and preventing the flux from dropping off into the fire; but just before the pieces are ready to be removed from the fire turn the scarf down, and when taking out to weld, tap lightly, to shake off the dirt, but not hard enough to remove the flux. By keeping a clean fire a smith should have no trouble in welding any kind of weldable steel or iron.

We will now consider the different kinds of welds: Scarf, butt, lap, cleft and jump welds.

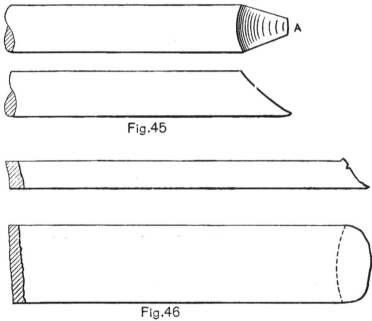

Fig. 45

Fig. 46

Correctly formed Scarfs for Welding.

Scarf Weld.—If about to make a scarf weld with round stock, the scarf should be made narrow, as at *A*, Fig. 45, which shows the top and side view of the same piece. Some smiths make the scarf wider than the bar to be welded, which is not a correct method, since in striking on the edge of the weld the thin part of the scarf will pucker and cool so quickly that another heat is necessary. Then the dirt will get in and a bad weld is the result, and especially so if flat stock is being welded. Flat stock should be scarfed about as in Fig. 46, which shows the edge and bottom views of a piece of stock.

Two Heats to Make a Weld.—A remedy for this useless waste of time and poor work in making a scarf weld is to locate the pieces with respect to the anvil as

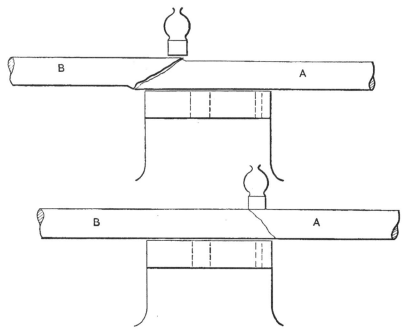

Fig. 47. Making a Scarf Weld.

in the two views in Fig. 47. When the helper takes the piece from the fire, he usually places it so the thin part of the scarf is at about the center of the anvil. The anvil chills the thin part of the scarf and by the time the smith sticks the scarf on top and turns the piece over to weld the other side he finds he must take another heat; and before he does so he will probably dig his fire all to pieces to start it up brighter, thinking, of course, that the fire was to blame. Now, if the helper, in taking the piece *A* from the fire, will place it as shown in the upper view, Fig. 47, so that the point of

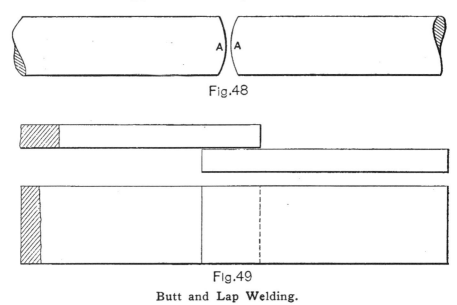

Fig. 48

Fig. 49
Butt and Lap Welding.

the bottom scarf will not touch the anvil, and the smith stick the top scarf with his hammer, and then

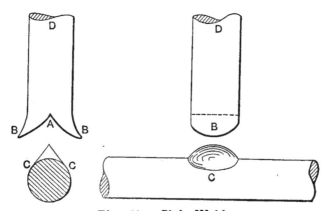

Fig. 50. Cleft Weld.

turn the piece over, placing as in the lower view, and both smith and helper go after it lively, there will be no trouble about welding.

Butt Weld.—The butt weld is the strongest weld for heavy, round stock, and is the easiest weld to make. The ends of the pieces to be welded should be slightly convex, as at AA, Fig. 48. This allows the centers to come together and also permits the slag to escape. The hammering done on the ends of the bar in making this weld supplies all the jumping up required; for if jumped up too much before welding the centers are liable to be drawn apart when drawing the piece down to size.

Lap Weld.—The most important points to observe in making a lap weld are to strike first at the center (X in Fig. 49) and then work outward from the center in all four directions. This will work out the slag and leave a clean, strong weld.

Cleft Weld.—In Fig. 50 is an end and a side view of a cleft or T-weld made with round stock. It is important to make it as strong as possible at the center A, in order to make a good weld, so we first stick lips BB to the lower bar at points CC. One must strike a few heavy, rapid blows on top of part D, to drive out the slag. If slag is not removed we cannot get a good weld of any kind. Bulge C is made by using the ball pene of the hammer, if the piece to be welded is small. If it is a large piece a ball pene top tool is used and the metal worked out from the bar gradually.

Jump Welding.—In Fig. 51 are two views to illustrate jump welding. The point A should be so shaped that it will touch part B first, before the edges come in contact. By looking out for this and striking heavy blows on the end of piece C all the slag can be driven out from between the two pieces being welded. When the pieces are well stuck together, use the fuller around shoulder *D.* Fig. 52 shows the jump weld job when completed. It will be noticed that the stem *C* is on the opposite side from where it is in Fig. 51, making it necessary to show the stem by a dotted circle

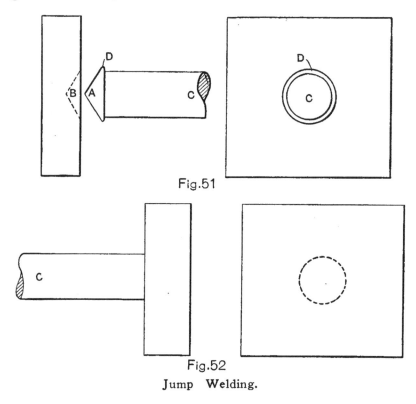

Jump Welding.

instead of by a circle drawn with a full line as in Fig. 51. This explanation is just to help out on the drawing question.

Welding a Ring.—In making a ring, large or small, most blacksmiths heat a bar for a distance of about 12 inches and bend it over the horn of the anvil; then heat a little more and bend and fit to a circle on a board or face plate. After they have bent what they think is enough stock to complete the circle and weld they cut the bent piece off the bar and scarf and weld it. After welding the ring it may perhaps be necessary to draw

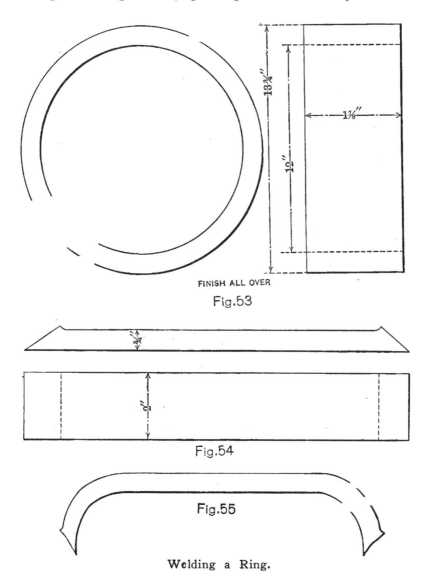

FINISH ALL OVER
Fig.53

Fig.54

Fig.55

Welding a Ring.

it until there is no stock left for the lathe hand to take off; or if the ring is too big it must be cut and welded again.

I am now going to give a simple rule for welding a ring that is worth more than five times the price of this book. It is a rule for making a ring no matter how large or small. I have made large rings from two-inch square stock for the smoke boxes of locomotives and never had anything more to do to them when finished than true them up and lay in a pile ready for drilling. The rule is as follows:

Add the thickness of the material to the inside diameter; multiply by 22; and divide by 7.

This furnishes jump-up stock and the amount for welding in all cases. For instance, if I had to make a ring to finish 1⅞ by ⅞ inch, 13¾ inches outside diameter and 12 inches inside diameter, as in Fig. 53, I would use stock 1 by 2 inches. I would add the thickness of the stock, which is one inch, to the inside diameter, which is 12 inches, making 13 inches. Multiplying this by 22 gives 286 and dividing by 7 gives 40 6/7 inches. I would cut off a piece of this length, jump up the ends and scarf each end on the *same side* as in Fig. 54; then bend backward, as in Fig. 55. Any blacksmith who tries this method of making rings will never make them in the old-fashioned way again. Some smiths require more for welding than others, in which case some may have to cut off more to allow for this. Of course the way to do is to try it out for yourself and become convinced as to the operation of the rule and determine whether allowance for more stock must be made in your case. The rule is adapted for stock of any shape, as round, flat, square, etc.

Welding Solid Ends in Pipe.—For the benefit of blacksmiths who are up against the steel-pipe roll welding proposition, I will give my experience along that line. I have welded steel pipe in sizes from 1 inch to 3 inches for over eight years, making hollow rolls with solid ends used in the construction of printing presses.

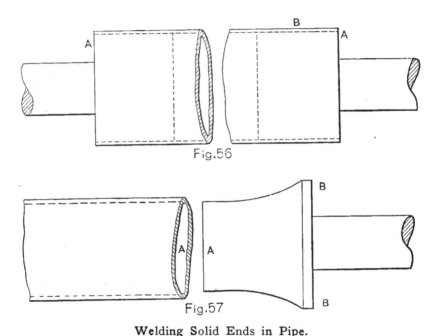

Welding Solid Ends in Pipe.

The rolls are turned and finished to a certain size after they leave the blacksmith. During this period that I was on printing press work I have seen a number of smiths who were not successful in welding up their rolls. Some would try to drive a hot plug into a cold pipe; others would try to drive a cold plug into a hot pipe and when they tried to weld the pipe it would pucker up and the job was thrown into the scrap. I have seen the plug fall out in the lathe after turning off stock at the end of the roll (*A* in Fig. 56) and I have also seen the rolls scrapped on account of being too small at point *B*.

Of the thousands of pipe rolls I have welded I have been fortunate enough not to lose one, and my method was as follows: First I had the rolls cut off to the finished length and no more. Then I made the plug as shown in Fig. 57, nearly ½ inch larger at *B* than

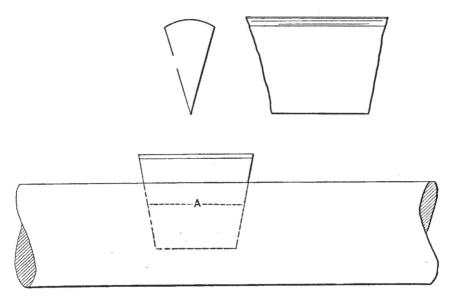

Fig. 58. The use of a "Dutchman."

at *A*. The plug would just start in the pipe at *A*. I heated both pipe and plug as hot as I dared to and drove together as quickly as possible and returned to the fire to take a welding heat. When welding I would get around the pipe quickly. You cannot get a good welding heat on a plug after it is driven into a pipe cold, or at a low heat. By leaving the plug large at *B B* it gives you something to hammer on and draws out long enough to furnish stock for turning to finished length. Fig. 56 shows a pipe roll after being welded.

The Use of "Dutchmen."—The small piece shown in Fig. 58 is commonly called a "dutchman" and is ordinarily used in covering up a bad spot, or to help out a poor weld. It is, however, used to good advantage in lengthening a bar, as it keeps it up to size and at the same time gives the desired extra length and saves cutting and welding. A dutchman is usually made by thinning down one side of a round bar, and should be wider at the top than at the bottom. If you want to lengthen a large bar heat a spot on the bar and drive in the hot chisel; then drive the cold dutchman in the hot bar, as shown at A; then take a welding heat and you will be surprised at what a nice job can be done in a short time. The dutchman should not be driven in more than one quarter the depth of the bar. If the bar needs drawing very much you can use a dutchman on each side.

CHAPTER II.

TOOL FORGING.

Systematic Arrangement of Work.—It is seldom that the foremen in machine shops give sufficient attention to the systematic handling of the tool work that is sent to the blacksmith shop. A man will leave his lathe, go to the blacksmith shop, and wait for a tool to be dressed or tempered, as the case may be, when perhaps he could have found what was wanted on the floor of his own department. I have seen as many as six lathe men standing around the tool dresser waiting for tools and as many as one hundred worn out lathe tools that needed dressing lying around on the floor under the feet of the workmen. A smith cannot do good work under these conditions, and although he may be the best workman in the country, he will be all at sea under such circumstances.

A plan that I have followed for the systematic handling of tool work, and have had adopted in different shops where I have been employed, is to have cupboards in different departments of the machine shop, as illustrated in Fig. 1. The foreman can keep his coat and hat in the part marked *A,* and tools to be dressed or tempered are to be placed on the shelves marked *B*. Tools that have been dressed and tempered are to be placed on the shelves marked *C*. This latter part is kept locked except when taking out tools to give to machine hands, or when putting in tools that come from the smith shop. There are two keys for this, one for the foreman and one for the boy who takes the tools from the compartments *B* to be dressed or tempered, and replaces when finished to shelves *C*. In this way the different kinds of tools can be separated, each kind having a shelf. If the door is not kept locked, one workman may have more than his share of tools, while his neighbor will not have enough.

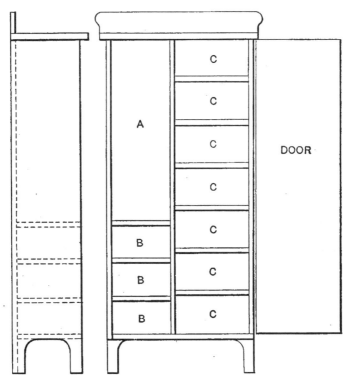

Fig. 1. Tool Cupboard for Machine Shop.

The section at the left has no door and any machinist having tools that need dressing can put them on the shelves himself, which saves time and gives satisfaction to all concerned.

To go with the cupboards located in the machine shop, the blacksmithing department should have a box similar to that illustrated in Fig. 2. It should be divided into as many spaces as there are departments in the factory, and placed near the tool dresser's place of business. Each space should be marked with the department number (see 1, 2, 3, etc., in Fig. 2) and all tools and small parts brought into the blacksmith shop should be deposited in the proper compartment. With this arrangement the boy delivering the tools cannot make a mistake, and the tool dresser has perfect freedom to work to the best advantage, which is not the case where a man goes into the blacksmith shop with a tool and wants to know how long before he can have it, saying that he cannot do anything until he gets it, and so on; and then before he leaves another arrives,

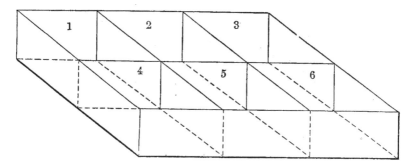

Fig. 2. Tool Tray for Smith Shop.

and then another until, as I have said before, there may be six men standing around the tool dresser talking, arguing and nearly quarrelling. The system described above obviates this, and in every instance where I have brought it to the attention of those concerned it has been approved and adopted.

Care in Heating.—One cause of poor results in tool work lies in the smith's carelessness in heating steel, going to the extremes by heating it either too fast or too slowly. I have known smiths to have a tool in the fire, heating it for dressing, when some one would come along looking for a chance to kill time. The two would get into a conversation, the smith leaving the blast on a little, until the tool became too hot; when he would remove it from the fire. After a little while he would put the tool back and heat it to a white heat, then remove it again, and repeat this operation perhaps half a dozen times before his company would leave. But tools handled in this way may as well be thrown away as to be sent into the machine shop, as they will not stand up to their work.

Putting a tool in the fire and not turning it over at all is very bad practice. A smith should put the tool or steel in the fire, cover it and roll it over repeatedly. He should watch it carefully, and heat it slowly. When it reaches the proper heat he should go at his work in earnest, striking heavy blows until it begins to cool off, after which lighter blows will do, depending of course, upon the kind of tool.

A smith engaged in tool work should not do any welding if he can avoid it. If not kept busy on tools he should be at some work that requires a low heat. Welding will cause him to become careless and he will spoil the tools in spite of himself.

Patterns.—A foreman smith should have patterns of all kinds of lathe tools, made from pine. In this way he can have any one of the blacksmiths make tools, thus teaching them to take the tool dresser's place should the tool dresser leave. I have known a foreman to take a worn-out lathe tool to the smith to be redressed and neither one understood what was wanted. The order system and the wooden patterns would do away with this trouble. When a smith makes a new tool he should stamp on it the kind or grade of steel used.

Bevel Set.—The tool dresser's most important assistant in making a success of tool work is the bevel set hammer, shown in Fig. 19, Chapter I. It makes one straight side on all his work and leaves a fillet, instead of a sharp corner to crack in tempering. It can be used in nearly all kinds of tool work shown in this book. It is much better than a fuller, which makes a rough looking job at the best.

The next important point is to have about three inches on each side of the anvil rounded to about ⅛-inch radius. This furnishes the necessary fillet so essential in all machine tool work. The balance of the anvil face can have sharp corners which are very nice for some kinds of machine steel forgings.

Directions will now be given for forging the more common tools used in machine tool and hand work that the blacksmith has to deal with, and in Chapter III. the hardening and tempering of these tools will be taken up.

Cold Chisels require more attention than is given them by the majority of blacksmiths, who either do not know or do not care. There is no class of tools in any plant, large or small, that we hear more grumbling about than the cold chisel. Very frequently a machinist buys his own steel and has a chisel made by some one other than the one used to doing his work, in the hope of getting a good tool, when the steel in the factory where he works is no doubt just as good as that which he buys and the factory smith can make just as good a chisel by being a little careful. The trouble is that the tool dresser has a lot of work of this kind to do, and, thinking that anything will answer for a cold chisel, he becomes careless. He does not think of the inconvenience to which the user is put if he has to produce results and make a good showing with a poor cold chisel.

When making a chisel it should be heated slowly so as to heat all through and when taken out of the fire it should be worked with heavy blows while it is hot. Hammer on the edge as little as possible and always

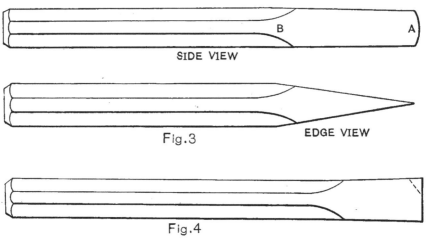

Right and Wrong Shapes for Cold Chisels.

strike the final blows on the flat part of the tool. When it commences to cool it should be reheated and should always be drawn down square; then it should be widened to the proper width, which should never be as wide as the bar it is made from. Fig. 3 shows the proper idea for a flat cold chisel and Fig. 4 the wrong way to leave a chisel. The end A should be ground a little rounding and should not be as wide as part B.

If the end is left wider than the bar, it will nearly always break at one corner, as shown by the dotted line in Fig. 4.

Cape Chisels, Grooving Chisels, Etc.—A cape chisel Fig. 5, should be drawn down square at first, then fullered as in Fig. 6, then forged to shape as in Fig. 5. All chisels and punches should first be drawn down square before forging to shape, as by so doing the center of the steel comes out as at *A*, Fig. 6. Another important thing about chisels is always to cut plenty off the drawn out end. Make a cut on each side and after the chisel is tempered, break off. (See dotted line at *B,* Fig. 5.)

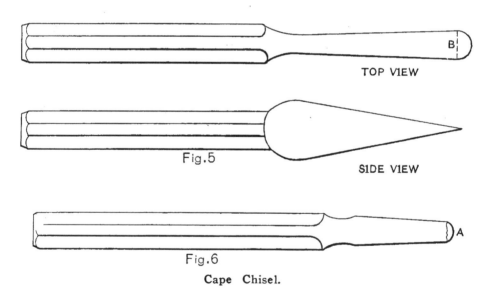

Cape Chisel.

A round-nose chisel is made about the same as a cape chisel, only a swage, as shown in Fig. 10, is used to make one side round. The chisel can be made in different sizes by having different sized grooves in the swage. The grooves should be deeper on the end.

TOOL FORGING 47

nearest the smith, and run almost out at the other side (see dotted line in the figure) to get the best results.

In Fig. 7 is a common round-nose chisel.

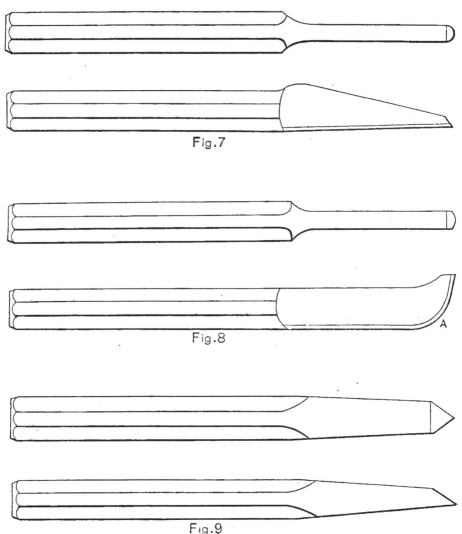

Round Nose, Grooving and Diamond Point Chisels.

Fig. 8 shows a grooving chisel for making oil grooves in bearings. Some smiths bend the common round-nose chisel to the shape of Fig. 8 at *A,* but it will either bend or break off.

Fig. 9 shows a diamond point chisel and Fig. 11 the side and end view of the swage for making such tools as diamond point chisels and other three-cornered pieces.

Screw Driver.—In Fig. 12 is a screw driver that will be appreciated in tool room or machine shop. It can be made from round or flat steel. I have sometimes

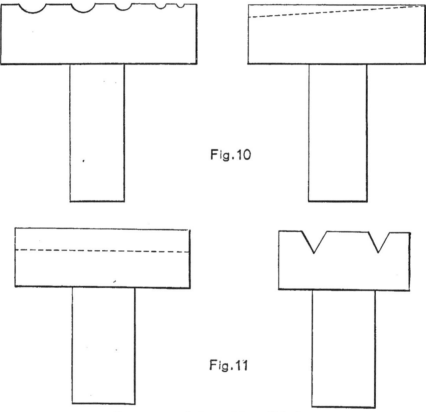

Swages used in making Chisels.

made them from octagon steel. This is a very handy tool.

Tools from Files.—Sometimes a smith is called on to make small tools, such as scrapers, small chisels, punches, etc., from old files. In such cases the files should always be ground smooth before making a tool of any kind from them, as the nicks and grooves will pound into the steel, causing the steel to crack in hardening.

Cutting-off Tool.—The first lathe tool that we will take up will be the cutting-off or parting tool. Fig. 13 shows how it should be made. Always use the set hammer shown in Fig. 19, Chapter I., then flatten down as in Fig. 14, keeping the top side straight and letting

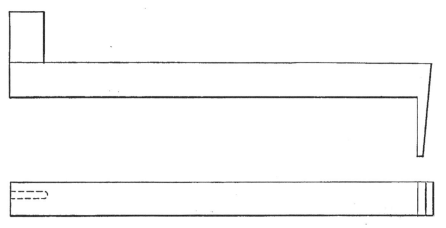

Fig. 12. A Handy Screw Driver.

the surplus stock go down on the lower edge of the tool. Do not hammer on the edge, but trim off as shown by the dotted lines in Fig. 15. Some cutting-off tools are used straight and some are bent. (See Figs. 16 and 17.)

Fig. 18 shows two views of a cutting-off tool that is a favorite with machinists who have used it. It is made by offsetting the stock before forging the blade. This offsetting can best be done under the dies of the steam hammer if one is at hand, as illustrated in Figs. 19

50   *THE BLACKSMITH'S GUIDE*

and 20. Here *A* is the piece to be offset, *CC* are two tool steel blocks so placed as to perform the offsetting,

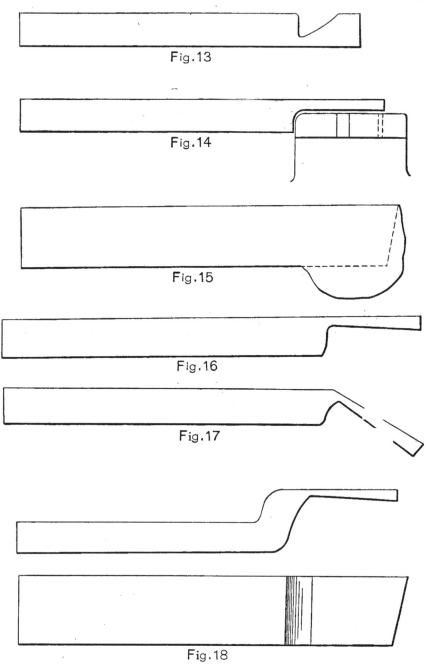

Making Cutting-off Tools.

and *BB* are the steam hammer dies. Fig. 20 shows the piece after being offset. Great care must be taken when doing this, as there is danger of either of the blocks *CC* shooting out if the piece about to be offset is not properly heated. Once I had the painful ex-

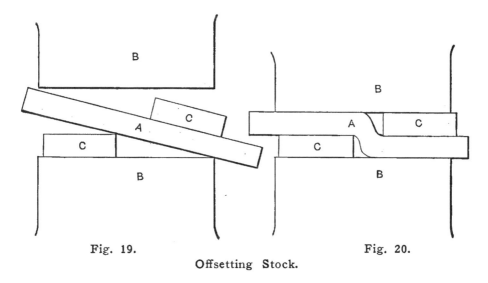

Fig. 19. Fig. 20.
Offsetting Stock.

perience of being struck on the knee by one of the pieces shooting out, caused by the carelessness of the blacksmith who was inexperienced in tool steel work and was trying to offset a piece of lathe tool steel at too low a heat. There is no danger if the piece is properly heated.

Side Tools.—In Fig. 21 is a top and side view of a side tool. It should not be pounded on its edges but should first be formed to shape on the lower edge, *A,* causing the metal to flow toward the top of the tool as indicated in Fig. 22. The edges should then be trimmed as shown by the dotted lines in Fig. 22. The cutting edge should be a little higher at point marked *E* than it is at heel marked F. When cutting, start chisel at *C* and incline the chisel at an angle so as to run out at *D*. The same clearance should be given the end of the tool as at the top.

Figs. 23 and 24 still further show the process of forming the tool on the anvil. It will be noticed in Fig. 23 that the tool is placed on the anvil with the top side down and that the bottom of the tool is being

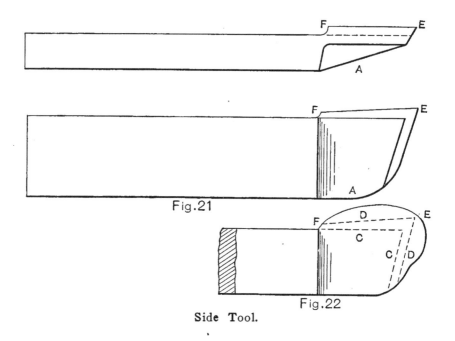

Side Tool.

brought to shape by the blows of the hammer. In Fig. 24 the edge of the tool is being drawn down thin on the face of the anvil, leaving the back of the tool at *A* as thick as the bar it is made from. Side tools are made right and left hand as the case may be. The one in Fig. 21 is a right-hand tool.

Boring Tool.—Next is the boring tool. After using the bevel set, Fig. 25, draw out as in Fig. 26. The lower view, Fig. 26, shows the boring tool after the lip

TOOL FORGING

has been turned. Fig. 27 shows the process of turning the lip over the round part on the anvil face. The lip on the boring tool should be made in the shape of a hook for turning tool steel, machine steel or wrought iron, as in Fig. 28. The hook can be formed by using

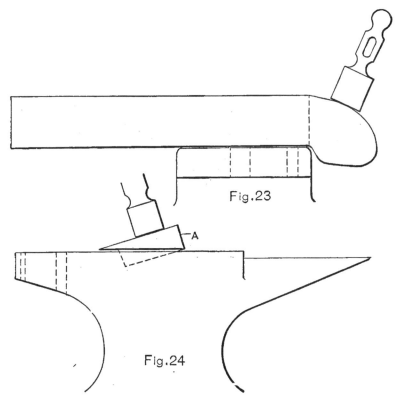

Forging a Side Tool.

the bottom fuller, as in Fig. 29. If boring tool is for brass the lip should have no hook, but should slant slightly from heel to point. Figs. 30 and 31 show the cutting parts for a brass-cutting tool.

The advantage of having round corners along part of the front and back edges of the anvil is well illustrated in the case of this tool, since they enable the tool to be forged with fillets instead of sharp corners at points *A* and *B,* Fig. 26, making it much stronger and less liable to crack in hardening.

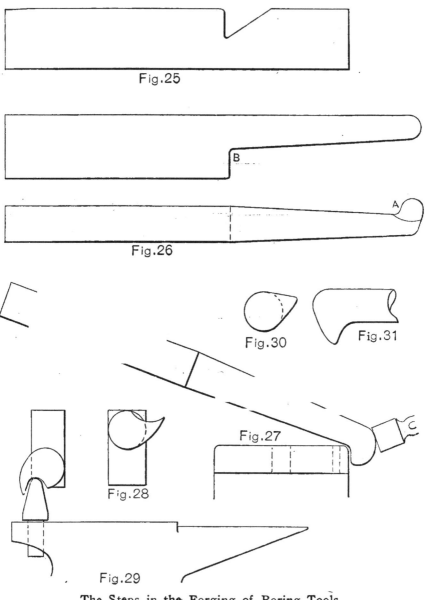

The Steps in the Forging of Boring Tools.

After a boring tool has been in use for some time and has been to the blacksmith to be dressed a number of times, it usually requires lengthening. In doing this, do not attempt to cut out a piece, as in Fig. 33, as you are almost sure to get a cold shut. Just take the bevel set and drive down as shown at *A,* Fig. 32, cut off the

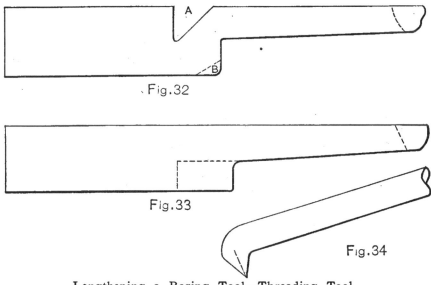

Lengthening a Boring Tool—Threading Tool.

corner at *B,* and then draw out. If you have to lengthen an almost new tool the operation will throw the lip upside down, but by twisting the round part of the tool it will put it back where it belongs.

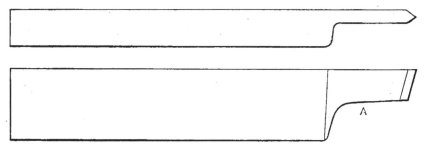

Fig. 35. Threading Tool.

Threading Tools.—An inside threading tool is made about the same as a boring tool, only the point is cut

off differently, as shown in Fig. 34. An outside threading tool is made about the same as a cut-off tool, only it is cut to a point instead of square. (See Fig. 35.) It can be cut out underneath as at *A,* Fig. 35, which leaves less grinding for the lathe hand, and a threading tool, not having to endure much strain, need not be very wide, like a cut-off tool, which has hard work to perform.

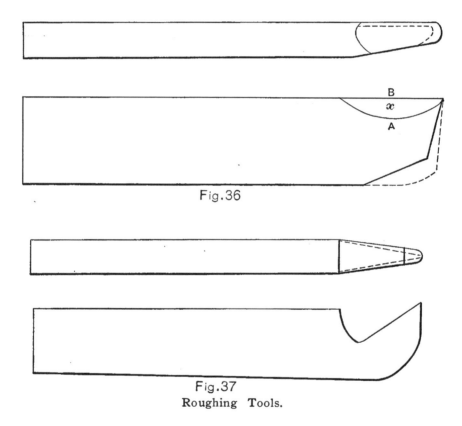

Fig. 36

Fig. 37
Roughing Tools.

Roughing Tool.—In Fig. 36 is a top and side view of a roughing tool which is very much in demand and is one of the tools that is used both by lathe and planer hands, and is easy to make. Just draw out like a common round-nose tool, and then with a half-round or C chisel cut out the portion marked *x*. Start the chisel at *A*, and slant enough to let the chisel run out at *B*.

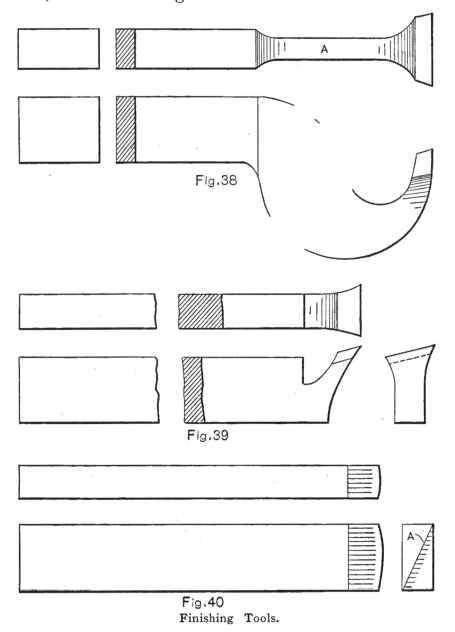

Fig. 38

Fig. 39

Fig. 40
Finishing Tools.

If the tool is for a planer it should be forged the same as shown by the dotted lines, giving less clearance than where used for lathe work. These tools are made both right and left hand. The one shown feeds toward the left and should be straight along the left side. The dotted line in the upper view, Fig. 36, indicates clearance, if tool is for planer or shaper. All planer and shaper tools should have less clearance than lathe tools, for both roughing and finishing.

Fig. 37 shows a roughing tool for shapers, and it can be ground so as to be used for either right hand or left hand. The dotted line in the upper view indicates the amount of clearance the tool should have.

Finishing Tools.—For cast iron there is no better tool for finishing than the one shown in Fig. 38. It should be made quite thin and wide at A and the cutting edge should not be more than $3/8$ inch above the bottom of the bar and should be tempered quite hard. There is no tool that will make a better finishing cut on cast iron than this one.

Fig. 39 illustrates a lathe finishing tool which works well on all kinds of steel and cast iron. This tool should be tempered very hard as it has little to take off and stands small chance of breaking.

In Fig. 40 are views of a finishing tool for steel on planer and shaper work, and which is used both right and left. This makes a nice finishing cut on all kinds of steel and, like other finishing tools, should be tempered very hard.

Centering Tools.—In Fig. 41 is shown a centering tool which differs from the ordinary tool of this description in having the end bent at an angle of 45 degrees. It is the only centering tool the machinist will use after once trying it. A straight centering tool

TOOL FORGING

is shown in Fig. 42, and in Fig. 43 is a centering tool made out of round steel, for turret lathes. In making these tools they should be flattened out and trimmed off the same as a long, slender drill. Never try to hammer to shape; always trim off, as it makes much better tools. They should not be tempered very hard, as they have a delicate task to perform.

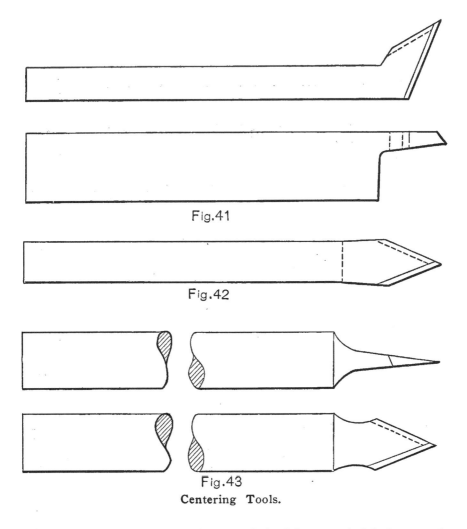

Fig. 41

Fig. 42

Fig. 43
Centering Tools.

Diamond Point.—The tool holder and high-speed steels have driven the diamond point lathe and planer tool almost out of commission, although it can still be found in use in some shops. To forge a diamond point, drive the fuller about half way through the bar as at *A* in No. 1, Fig. 44. Hold on anvil as shown in No. 2, striking heavy blows on corner *A,* and then turn so corner *B* will come up, as in No. 3, and strike heavy blows. Repeat this operation until you have the end

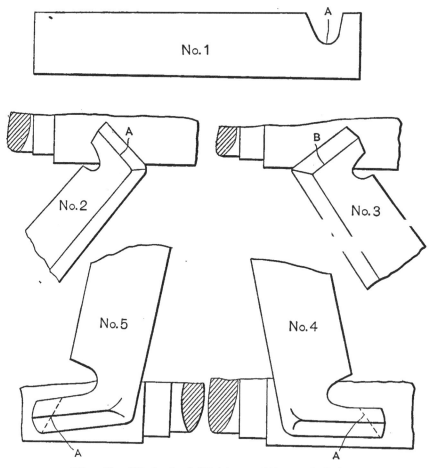

Fig. 44. Method of Making a Diamond Point.

drawn square. If for a right-hand tool, place on the anvil, as in No. 4, holding the tongs in the left hand and hot chisel in the right hand, and start the chisel at

A, with top of chisel leaning well towards you, thus giving clearance to the tool. If for a left-hand tool proceed as in No. 5, holding tongs in right hand and chisel in left hand, starting chisel at *A.*

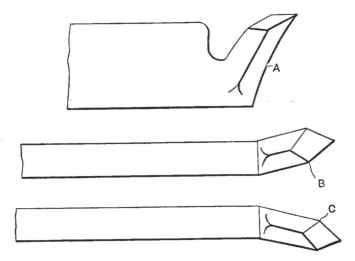

Fig. 45. Diamond Point for Lathe.

If the diamond point is to be a right-hand lathe tool, it should be slightly curved as at *A,* Fig. 45 and bent a little as at *B.* If for a left-hand lathe tool, it

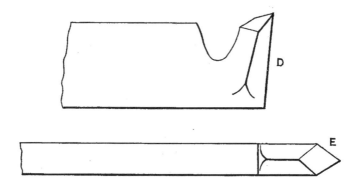

Fig. 46. Diamond Point for Planer.

should be bent the reverse, as at *C.* If the tool is for a planer or shaper, it should have very little clearance as at *D,* Fig. 46, and should be left straight as at *E.*

Round-Nose Tool.—The common round-nose tool, shown in Fig. 47, is used to finish fillets in work and should slant slightly from *B* to *A*, making the point *B* a little higher than the top of the bar.

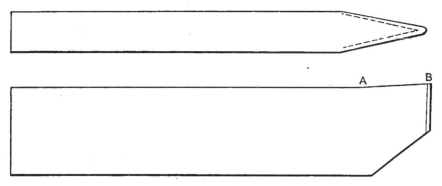

Fig. 47. Round-nose Tool.

Brass Turning Tool.—A V-tool, which is the best tool for brass turning, is shown in Fig. 48. It should be so made as to slope down slightly from *A* to *B*, and should be tempered very hard.

Fig. 48. Brass-cutting Tool.

Rock Drill.—While not to be classed with machine shop tools, the making of a rock drill is a tool steel job that a blacksmith often has to do. Fig. 49 shows how to make a four-lip rock drill, the fastest working rock

TOOL FORGING

drill to be had and one that is easily made. After drawing the shank *A* down to the desired size, take the top and bottom fuller (about a half-inch fuller, if the drill is to be standard size), and form as at *B* in the two upper views. Then draw out the parts *CC,* etc., to the shape shown in the two lower views, and trim off with the chisel, cutting the edge as shown. This drill should be tempered the same as cold chisels, described in the next chapter.

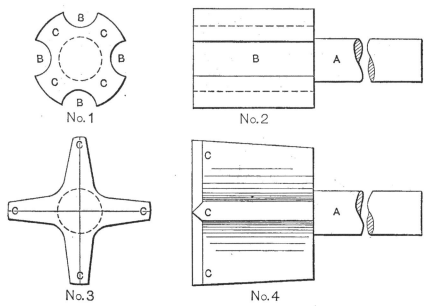

Fig. 49. Four-lip Rock Drill.

To Make Tool Steel Rings or Dies without Welding.—All workers in tool steel know that after a piece of tool steel has been welded it will not have the same fine grain as before. In spite of all we can do the part welded will have a coarse grain and much strain will cause it to crack so that in tempering it there is liable to be more or less trouble from this cause. In making dies, therefore, it is always advisable to avoid welding, which can be done as follows:

Select a piece of steel of a suitable quality and of the right thickness for making the die. The piece

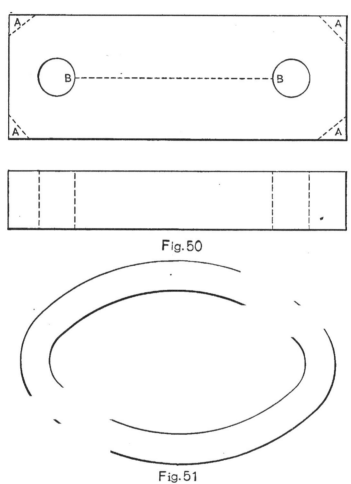

Fig. 50

Fig. 51
Tool Steel Ring Made without Welding.

should be one half the length and double the width that would be used in welding a ring according to the directions already given. Fig. 50 shows the piece of steel and the method of opening it up for the die. First cut

TOOL FORGING 65

off the corners *AA*. Then punch holes *BB* far enough from the ends to leave stock to clean up. Split along the dotted line *BB* and open up as in Fig. 51 and form

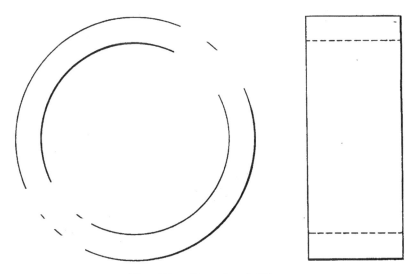

Fig. 52. Completed Ring.

a perfect ring. This method should be followed in making either tool steel rings or dies. It is just as easy as welding and makes a much better job.

CHAPTER III.

HARDENING AND TEMPERING.

When a piece of tool steel is treated to make it harder, we speak of the process as "hardening" and when a piece of tool steel, already very hard, is treated to make it suitable for a cutting tool by drawing the temper, the process is called "tempering." When a piece is heated to the correct temperature and dipped, according to directions that will be given, it is usually not necessary to draw the temper and the process is properly called hardening, although the treatment of steel for cutting tools is frequently called tempering, regardless of whether the temper has to be drawn or not. This is because it always used to be the custom to harden the steel and then draw the temper, the whole process finally coming to be called "tempering."

The art of hardening and tempering high-grade tools is understood by very few mechanics and the majority have had but little experience in tempering even a cold chisel. It is commonly thought that tools, bought in standard sizes and in large quantities by the factories throughout the country, and stamped with the trademark of some large firm, are O. K., and most mechanics would give a great deal to know how to harden tools the way these are hardened. I have known a machine shop foreman to send into the smithing department a lot of machine steel pieces to be annealed so they could be drilled or machined; and on investigation found the brand new drill that had been used on the pieces was too soft to drill anything except lead. After hardening the drill the pieces were drilled and machined with ease. The fact is that foremen, and men under them, take for granted that because a drill is a twist drill, and is bought, it must necessarily be all right, whereas oftentimes the tool is not all right and if it were made in the shop where the work is being done, the man who hardened it would get the blame, just as he ought to.

It is not my purpose in this book to explain how steel is made and what its composition is. I want, instead, to tell how the common blacksmith can give entire satisfaction in hardening and tempering, as carried on in the manufacturing plants of to-day. In the first place the tool steel used by the blacksmith is bought by the purchasing department, and when a tool comes into the smithing department to be hardened, it is up to the blacksmith to make the best of the situation and get the tool done as quickly as possible, without attempting to find out the percentage of carbon in the steel, or going into other scientific points.

In my experience on nearly all kinds of tools, I have found that each has to be hardened to suit the class of work it is intended to be used upon. For instance, I should not treat a tool for cutting brass the same as I would one for cutting machine steel. The actual cause of a great deal of trouble in all large plants is that the man who is to use the tool does not inform the one who is to do the hardening what is required of it; and if the tool fails the man using it blames the one who hardened it, and he, in turn, blames the steel.

Proper and Improper Heating.—About the worst possible practice for the smith is to have a dirty, shallow fire and to put a piece of tool steel in it, either to forge or harden. The air blowing on it ruins the steel, making it unfit for anything but scrap. Always have a good, deep, solid fire and heat tool steel with as little blast as time will allow, if heating for forging, and when heating to harden or temper, put in charcoal, and use no blast whatever. If these directions are followed you will be surprised at the demand there will be for your services.

The practice of the writer in hardening tools is to heat in a slow fire to the lowest possible temperature that will secure the desired result, and then dip, according to the directions given. This method does away with much unnecessary labor and avoids warping and cracking of valuable tools that is sure to occur where they are brought to too high a temperature, cooled, and then have the temper drawn back.

In order to emphasize the importance of proper and careful heating of tool steel, the micro-photographs in Figs. 1 and 2 are shown. Two bars of steel were selected, one of which was heated too hot for hardening and then dipped, while the other was heated slowly and dipped at a much lower temperature. These bars were then broken and the first showed a coarse, uneven grain and was useless for anything but scrap. Steel in this condition cannot be drawn to any color that will make it serviceable as a cutting tool.

The second bar had a fine, uniform grain throughout. The difference in the structure of the two bars was clearly visible to the naked eye, but to make it more evident to the reader, very much enlarged photographs were made of different points in the broken

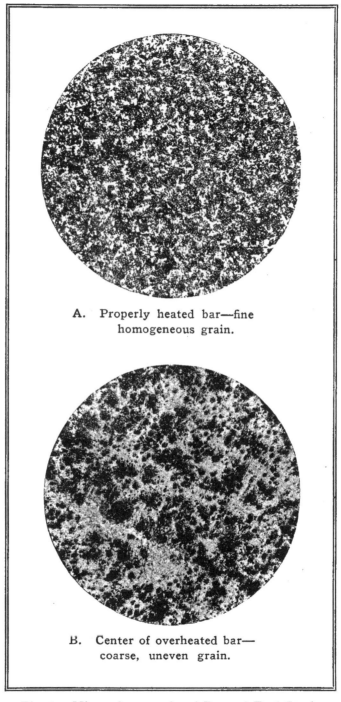

A. Properly heated bar—fine homogeneous grain.

B. Center of overheated bar—coarse, uneven grain.

Fig. 1. Micro-photographs of Bars of Tool Steel that were Properly and Improperly Heated.

ends of the two bars and are reproduced in the illustrations, Figs. 1 and 2.

The photograph at *A,* Fig. 1, is from the broken end of the bar that was properly heated. This bar is uniform throughout and has a fine, even grain, as indicated in the photograph. At *B,* Fig. 1, is a photograph taken at the center of the bar that was overheated, showing the coarse grain. Nearly the whole surface of the section is of this character, but at the outer edge, where the metal was suddenly cooled, the grain appears to be finer. The photograph *C,* Fig. 2, taken at this point shows the granular structure to have been completely destroyed. Photograph *D,* Fig. 2, represents the condition in the overheated bar at a point between the central section and the outer edge, where the most marked change in structure occurred.

If care is taken in heating steel, no attention need be given to drawing to any particular color afterwards. So many have followed the practice of drawing the temper of tools, however, that a few more remarks on this part of the subject will be in order.

Drawing the Temper.—I have read in books on tempering, "If you think you have the tool too hard, hold it over the fire to remove the internal strain." Now I claim that holding a tool over the fire will soften the cutting edges and leave the inside of the tool the same as before, especially if the tool happens to be a solid reamer or a tap. If it is a hollow mill or tap, and you want to "draw back," heat a bar of steel or iron $\frac{1}{16}$ inch smaller than the hole in the mill. Run it through the mill as in Fig. 3 and turn constantly. By putting a little oil on the mill you can see when it is

HARDENING AND TEMPERING 71

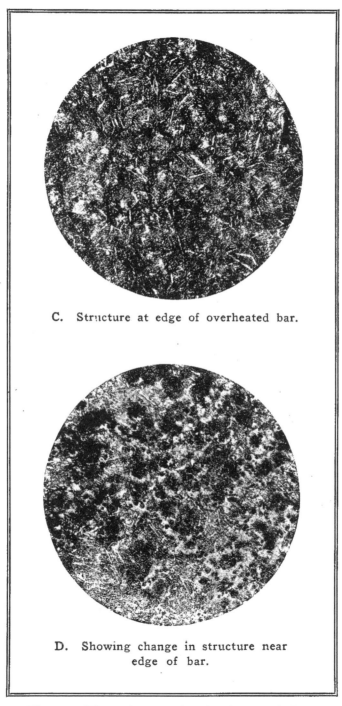

C. Structure at edge of overheated bar.

D. Showing change in structure near edge of bar.

Fig. 2. Micro-photographs showing Variation in Structure of Overheated Bar.

getting hot enough to remove. It is better to take it off a little too soon and let the mill cool off than to leave it until the last minute and then quench perfectly cold, as the former method leaves the tool much tougher. Tools hardened according to the instructions given in this book, however, will not need much drawing of this or any other kind.

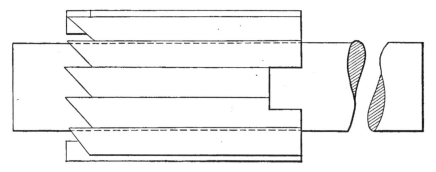

Fig. 3. Drawing the Temper of a Cutter.

Heating to a high heat and cooling entirely cold in water, as so often done where dependence is placed on drawing the temper to secure the right degree of hardness, always produces internal strains in tools of any kind. All mechanics have seen tools break in pieces before they were cleaned up for drawing back. A good way to convince yourself that such strains exist is to take two files. Break one and heat the other to a high heat. Cool off entirely cold and break it, and compare the grain of both files. You will find that the last one will resemble cast iron, and there will be as much difference in the appearance of the two as in the case of the steel bar shown in Figs. 1 and 2.

Again, take two tools dipped at slightly different temperatures and drawn to the same color. Will they have the same temper and be able to perform the same amount of work? I think not and yet this mistake in tempering is very often made. I can harden as many tools in five hours as anyone can in ten hours if he gets the tools too hard and has to "draw back," and tools that I have hardened by the methods described have records that are seldom equaled.

Charcoal Fire.—In my opinion nothing can equal charcoal for heating steel for hardening and tempering. Of course there are such things as lead baths, expensive furnaces, etc.; but how many shops have them, or will install them, even if the blacksmith should want them? His employers expect him to make use of the things at hand; and if he cannot, they will get someone who can. But almost any firm will buy a few hundred pounds of charcoal, especially when they see the results obtained from it when used as I am about to explain.

A smith can have a small home-made oven or forge near his regular forge and while the large tools are heating he can work at something else, because no blast is used in heating with charcoal, unless in a very great hurry, when a small blast may be turned on. I have put large hollow mills in such a fire, and been called away. On returning twenty minutes or half an hour later the mill would be ready to take out and dip.

Lead baths are very nice for heating large tools, but in my opinion cannot equal charcoal. With the lead bath you have to dip every tool at the same heat and if you get the lead to the heat that some tools require to be dipped at, it will not be right for the others.

Home-made Oven.—In Figs. 4 and 5 is a home-made oven or forge for use in hardening and tempering. It can be made from ⅛-inch sheet iron, and is supported by iron legs, built as shown, one at each of the four corners of the oven. Although shown with a stack and hood, at *A* and *B* respectively, neither of these is absolutely necessary. In fact the whole front *H*

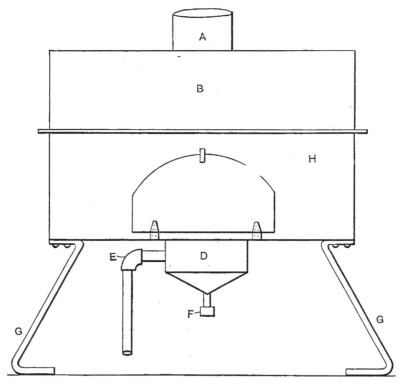

Fig. 4. Home-made Oven or Forge.

with door *C* could be done without. The side view, Fig. 5, shows the hood slightly raised. By having hinges on the back, you can tip the hood back out of the way, if desired, and use the forge in that way, since there is no smoke or gas from the charcoal. Sometimes, however, it is advisable to bank up the sides of a charcoal fire with smithing coal, as would be the case if the piece to be tempered were a long, large piece, so that the hood and stack are useful at times for carrying off smoke and gases. The tuyere *D* must be provided in every case. It is about eighteen inches long and six inches wide, and the whole top is drilled

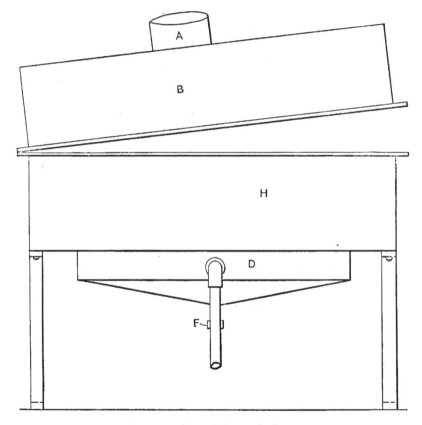

Fig. 5. Side View of Oven.

with ⅜-inch holes, about ⅜ of an inch apart. The bottom of tuyere should be slanting as indicated in the sketch, to allow dirt to escape through the blow-off pipe *F*.

The blow-off pipe has a cap screwed onto its lower end, which can be removed to clean out the tuyere.

When small tools are to be heated you can block off part of the tuyere by laying fire bricks at each end, giving by this means a narrow, short fire. The blast pipe E connects to the tuyere as shown.

The size of a forge or oven can be governed by the size of the plant and the amount of tempering generally done. It is a good plan to have quite a number of pieces on hand to be hardened before building a fire, as it takes only a short time to get the fire ready for business.

A still simpler forge is the open forge in Fig. 6, the illustration of which is from a photograph of the forge used by the writer in hardening most of the tools shown in this book. The forge is made from part of an old boiler and has neither hood nor stack.

Color Charts.—In the appendix will be found two charts to assist the smith in hardening and tempering. In the first chart, Fig. 1, will be seen colors to which different kinds of tools should be heated, and at which they should be dipped, for hardening. In determining the color of a heated tool it is better to have the room a little dark than very light. If the sun shines in or near the fire it is almost impossible to do justice to the tools. You should try to have about the same light at all times, which can be arranged by having shades on the window, if one is near. If the sun shines on tools in the fire they will be hotter than they appear to be by using the color chart as a guide, and if in the dark just the reverse will be true.

The first chart is numbered from 1 to 11 and indicates as nearly as possible the temperature at which tools hardened by the methods described in this book should be dipped. It will also serve in annealing tool steel.

No. 2 on the first chart indicates the correct color for dipping springs. This color will also do for small wood-working tools.

Fig. 6. Open Forge for Hardening and Tempering.

No. 3 is suitable for cold chisels, punches, small dies, small twist drills, small taps, etc.

No. 4 for larger taps and dies, larger punches, shears, large twist drills, etc.

No. 5 for all lathe and planer tools for cutting wrought iron and for tools of a delicate nature for cutting machine steel, such as small boring tools, threading, centering and side tools and cutting-off tools.

No. 6 for lathe tools for cutting machine steel and for planer and shaper tools; also for small shell reamers and pilots, broaches, etc.

No. 7 for lathe tools for cutting tool steel, cast iron, malleable iron, etc.; also for large hollow mills, large reamers and pilots of sizes over ½ inch.

No. 8 for finishing tools, or tools of different kinds for working brass castings; also hand scrapers and drill jig bushings.

No. 9 is the color at which tool steel should be worked. Steel heated to No. 9 forges easily and the smith should work lively until it cools to the heat indicated by No. 3 on the chart, and then reheat. When the tool is finished lay it down until it gets cool, when it can be reheated and hardened.

No. 10 is a welding heat for tool steel.

No. 11 is a welding heat for machine steel.

Certain tools like the side tool, which have a thin cutting edge, will harden at a lower heat than a more blunt tool, such as a roughing tool.

In case a tool has, by mistake, been heated too hot it is very wrong to swing it back and forth to cool it after it has been overheated. It is better to lay it down and let it cool off and then heat it over again.

Nos. 1, 2, and 3 of the second chart run from dark blue to purple; Nos. 4, 5, and 6 from purple to straw; No. 7 is light straw and No. 8 very light straw, which is about as hard as it is safe to get any tool for working purposes. I have not paid much attention to this kind of tempering for years, but it will be found convenient for beginners.

No. 1 on the second chart shows the color of steel when it is drawn to as soft a temper as can be used for tools. It is the color for springs and such tools as require a low temper, such as cold chisels and hand punches when used on soft machine steel.

No. 2 is suitable for cold chisels when used for chipping cast iron; also for the wood-working tools such as carpenters' planes, chisels, etc.

No. 3 for punches and dies used in punching thin metal in punch press work.

No. 4 for punches and dies when used on heavy metal in punch press work.

No. 5 for all lathe and planer tools for cutting wrought iron and for tools of a delicate nature for cutting machine steel, such as small boring tools, threading, centering and side tools and cutting-off tools.

No. 6 for lathe tools for cutting machine steel and for planer and shaper tools.

No. 7 for lathe tools for cutting tool steel, cast iron, malleable iron, etc.

No. 8 for finishing tools, or tools of different kinds for working brass castings; also hand scrapers.

Hardening Cold Chisels.—A cold chisel hardened in the usual way always comes back with the point bent up. The smith will generally heat about $3/4$ inch of the chisel on the end, then dip about $3/8$ inch of the point in the water; cool off entirely cold, then rub with a piece of grindstone or emery cloth. The color will run down rapidly. When it comes to a blue on the point he cools the whole chisel entirely cold and lets it go, thinking he has done pretty well. While a chisel tempered in this way is always too brittle on the extreme end, it will be too soft about $\frac{1}{4}$ inch from the end, causing the end to bend up as stated.

The way I treat cold chisels when I have a number of them to make is to forge them and let them cool off. After I have finished with the last one I begin again with the first one made, heating it slowly to No. 3 of the heat chart, for about $1\frac{1}{4}$ inches on the end, in the charcoal forge. I then dip the cutting end down slowly until the whole chisel is under water; but before it gets entirely cold I take it from the water and allow it to cool off on a bench or on the floor, as the case may be. If you cannot accomplish this at first, try it again. Once you get the required heat you will not forget it. In this as in all other cases salt water is to be used.

Another good way to treat chisels of all kinds is to heat as above and dip for about one inch in the water; take out and clean off, and let the color come slowly. If it comes too fast, check it a little by dipping in the water and taking out quickly; and if it does not come at all, caused by not having heat enough behind it, hold it over the fire, being careful not to get the point or cutting end near the fire. Hold the body of the chisel over the fire, and the heat will drive up the color slowly. It will show a mixed purple and blue about one inch long. By laying it down and not putting it in the water again you will have a chisel which will give the best of satisfaction.

Center Punches.—I have made center punches that were used on boiler work and were driven through solid boiler plate, thus forming a deep body of metal to be tapped out for bolts, and at the same time I found there were few smiths who could make these punches stand up for this class of work. A center punch should be made and tempered with the same care a cold chisel is.

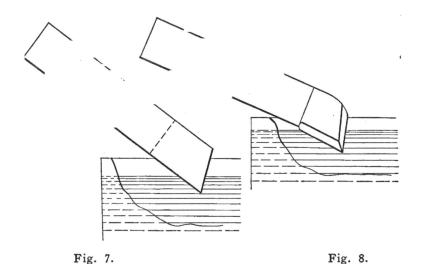

Fig. 7.   Fig. 8.

Lathe and Planer Tools.—Finishing tools require to be harder than roughing tools and should therefore be dipped at a little higher heat than the latter. It is necessary that roughing tools be tough as well as hard enough for the metal they are to be used on; and the same is true of tools of a delicate character, such as threading and cutting-off tools. Also, tools should be a little harder for cutting tool steel and cast iron than for machine steel and still harder for brass.

After forging a tool, let it cool off; then reheat the cutting point to the correct temperature for the work to be done, as already explained, and dip in salt water.

Remove from the water before entirely cold and allow it to cool off on the bench or floor.

What is desired is to have the steel tough except at the cutting edge, where it must be hard enough to cut the metal that the tool is to be used upon. In reheating, therefore, heat the point of the tool only, as stated, and in dipping hold the cutting edge under the water as indicated in Figs. 7 and 8, where the first is a cutting-off tool and the second a side tool; but do not immerse

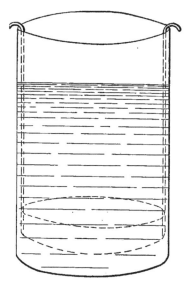

Fig. 9.

the whole end of the tool, as this will harden the part directly back of the cutting point and make a brittle tool.

If the body of the tool has been heated red hot through carelessness in reheating, it must, of course, be cooled off by dipping the whole tool; but the results will not be as good, since, if cooled off in the water, it will make the tool brittle, and if taken out of the bath before it is cool, enough heat may still remain in the steel to draw the temper of the cutting edge. In hardening tools, even of as simple a character as lathe tools, the smith must have a clear understanding of what it is desired to accomplish and then exercise care in bringing about the result.

Hardening Milling Cutters.—Have a tank or crock that will hold about 20 gallons. Fill with salt water in the proportion of about one half pint of common salt to the gallon of water. Also, there should be a

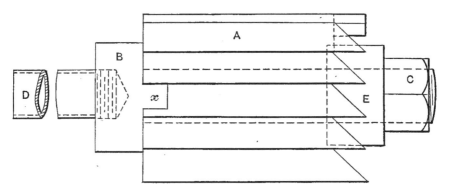

Fig. 10. Hollow Mill on Stud for Hardening.

second crock of the same size about two thirds full of fish oil, with a sieve suspended by wire rods over the edge of the crock, Fig. 9, so it will hang about six inches from the bottom.

Now if you are about to harden a hollow mill, as in Fig. 10, first have a steel stud turned up to fit the hole in the mill, and threaded on its outer end for the nut C. The other end B should be drilled and tapped out for a piece of half-inch pipe D, about 24 inches long, to be used to handle the mill with. The stud should also have two lugs or projections x to fit the slots in the end of the cutter. If the mill is to be heated in a furnace, no pipe handle will be required; but a furnace will scale the mill badly and tend to crack it and I prefer a charcoal fire instead.

Start a nice level fire in your oven or forge, place the mill on the charcoal and cover with charcoal lumps about the size of hens' eggs. Shut off the blast and let the mill heat. You will find it impossible to overheat it in this way. When the mill is at a good bright

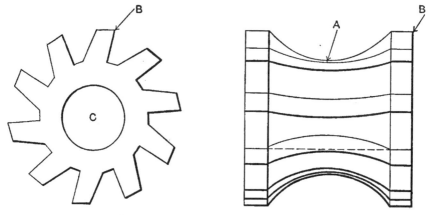

Fig. 11. Formed Milling Cutter.

heat, nearly as hot as the charcoal will heat it without the blast, take it out and dip it in the salt water crock. When the red is all gone and the vibration on the pipe is nearly gone, take the mill out of the water and put it in the crock containing the fish oil. Let it remain in the oil until it can be handled easily; then remove the nut and washer, take the mill off, put on another and repeat the operation. You can then send the mill into the tool room to be ground and put to work and it will do more work than if tempered in any other way that I am familiar with. This method hardens the outside of the mill only, and if at any time it is wished to use the mill on a larger arbor it can be reamed out without annealing, as would have to be done if the tool were hardened all through.

Hardening Formed Cutters.—A formed milling cutter shaped something like the one illustrated in Fig. 11 is a difficult tool to temper if done in the old

Fig. 12. Formed Cutter.

fashioned way of heating too hot, hardening too hard, and drawing back. Now if the utmost care is not taken this tool will be too hot at *B,* and not hot enough at *A,* and after hardening and cleaning up preparatory to drawing back, if it remains intact long enough, a bar of steel would be heated and put through cutter at *C,* and the color drawn in this way. The cutting edges at *A,* being nearest the hot bar, would naturally become soft before the cutting point or edges at B would be soft enough, B being the hottest when dipped and also the hardest when cold. You cannot obtain a uniform temper in a tool of this kind, when heated, hardened and drawn in this way.

The cutter shown in Fig. 12 is one of many tools of this nature, tempered successfully by the writer in the following way:

First arrange the cutter on a stud, as explained under the last heading, and cover with wood charcoal. Let it heat without much blast. If you use any blast in heating this tool you must roll the tool over once in a while, but if you let it heat with charcoal without the blast, the tool will heat uniformly, without any attention other than watching. Now have the water very cold and salt the same according to directions already given. When the tool becomes a clear red, not too dark a red, dip in the cold salt water. When the red has all disappeared and vibration on the pipe handle become less, remove to the oil and let it remain there until it is cold enough to handle with the bare hands. Now in dipping a tool of this kind or of any kind, do not move it sidewise, for by so doing you are entering cold water with one side of the tool and warm water is following up the other side, and you will be very apt to have a hard and soft side on a tool.

We all know that warm water comes quickly to the top of bath. So when dipping a tool, start it down straight, and go slowly to the bottom of the tank, can, pail or crock—whatever you are using. This is the only possible way to get a uniform hardness. The milling cutter shown in the half-tone referred to was used in milling tool steel pieces. It required 50 per cent less grinding than any standard tool bought in the market and used in the same department on the same grade of steel, but different shaped pieces.

If about to adopt this method of hardening you will find it convenient to have a drip pan with sieve same as shown in Fig. 13. Have the pan a little higher at the end marked *X*, farthest from the oil tank *A*.

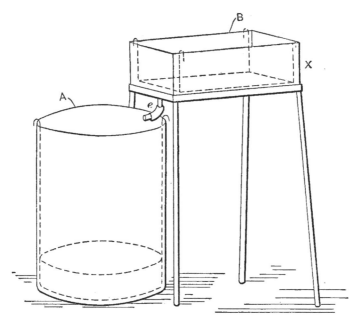

Fig. 13. Drip Pan and Sieve.

Then, when removing the pieces from the oil tank, place them in pan *B*. The oil will drip off and return to tank *A* through pipe *e* in the lower end of the pan. You will find this a very convenient and clean method.

Hardening a Thin Cutter.—To harden a milling cutter about ¼ inch thick and 5 inches in diameter in a way to prevent warping and cracking, fix it as in Fig. 14, but in heating put on a little blast, instead of shutting it off, and turn the mill constantly until you get the proper low heat. Then dip as explained in connection with the treatment of the hollow mill, Fig. 10, and you will have a soft tool, except in the edge outside of the washers EE. The same ½-inch pipe will do for any number of studs. If you have a mill or tool of any kind that you cannot use a stud in, stuff with asbestos.

Additional matter upon the treatment of thin cutters will be found in Chapter V.

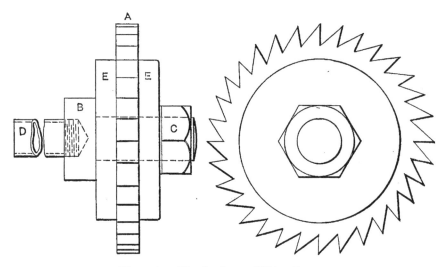

Fig. 14. Hardening a Thin Cutter.

The Treatment **of Reamers.**—In Fig. 15 appear a number of reamers just as they were taken from the oil. The longest one is 16 inches over all, with 12 inches cutting edge. The others are from 4 to 8 inches cutting edge. They were packed in charcoal inside of a 5-inch pipe, sealed up and put in a large casehardening furnace, and left there just one hour and twenty minutes. They were then taken out, one at a time, and dipped in cold salt water just long enough to harden the cutting edge; then removed to oil and left there until cold. They came out straight, and give the best of satisfaction.

Hardening by this method is known as "pack hardening" and more satisfactory results can many times

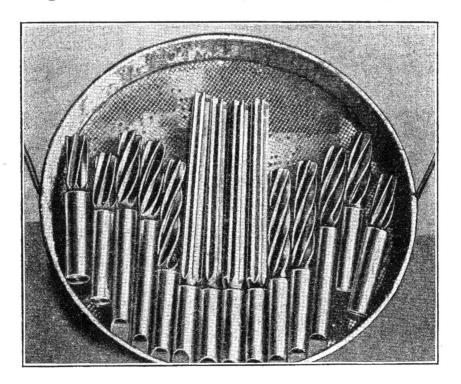

Fig. 15. Reamers with Cutting Edges Ranging from 4 to 12 inches long.

be secured in this way than by any other process, especially in the case of large and delicate work requiring uniform heating without injury to the metal in the cutting part of the tool, through decarbonization, or to the cutting edges themselves through internal strains. This subject will receive more attention in Chapter V.

If these reamers had been dipped in water at a much higher heat, and cooled off in the water, they would have sprung out of shape and would have had to be drawn back after cleaning them up. If you break a reamer or tool of any kind that has been hardened at

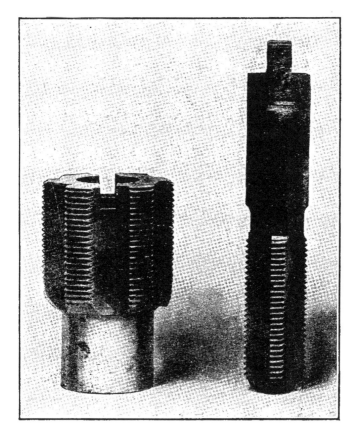

Fig. 16. 2 1-4" and 1" Taps.

a high heat, and compare the grain with a tool hardened as these reamers were, you will find the former is coarse grained and brittle, while the latter is tough and fine grained and will outlast anything hardened in the old fashioned way. Drawing to a color will not restore the grain but is liable to soften the edge and leave the inside brittle and at the least strain the tool will snap off.

Hardening Taps.—Of the two taps shown in Fig. 16, one is a 2¼-inch tap and the other is a one-inch tap. Both are of the same pitch—12 threads per inch. The total lengths of holes tapped in cast iron by these two taps was 7,500 linear inches, or 625 feet, and they were in fairly good shape at the end of this service. They were used both in piece work and day work by inexperienced hands. The taps are soft enough to be machined anywhere except at the teeth. The one-inch tap is solid and if required could be drilled down through the center the entire length from the top to about ¼ inch from the bottom. This makes the tool tough and allows it to spring; and it will not snap off the way so many tools do that are hardened all the way through. The 2¼-inch tap, being hollow, was arranged the same as a hollow mill, with the stud, etc., but was dipped at a much lower heat. Both taps were kept in the salt water just long enough to harden the teeth and then put in oil and left there until cold.

In Fig. 17 is shown a 1¼-inch pipe tap that, up to the time of photographing, had tapped 10,000 pieces of cast iron ¾ inch thick, some of them being very hard; in fact, some were so hard they had to be annealed in the furnace before they could be machined. This same tap had also tapped 10,000 pieces of malleable iron ¾ inch thick, all done on piece work. When photographed this tap was as good as new.

This tap was heated in a charcoal fire, very slowly, until the temperature corresponded to No. 4 on the heat chart. It was then dipped in cold salt water just long enough to harden the teeth, then thrown into the oil. It was allowed to stay there until cooled off, when it was returned to the machine shop and put to work on the pieces mentioned.

Treatment of Punches and Dies.—The $7/16$-inch punch and die shown in Fig. 18 have already punched

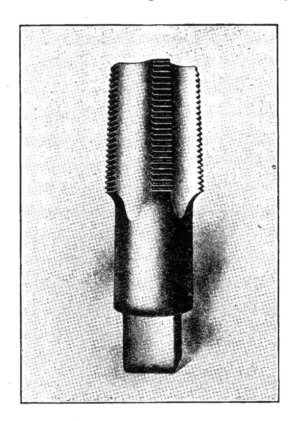

Fig. 17. 1 1-4" Pipe Tap.

100,000 holes in $1/4$-inch machine steel. This is 25,000 inches or a little over $3/8$ of a mile of solid metal, and when photographed they were as good as new and looked as if they might punch forever. The only hard part of the die is the bright part around the hole shown in center of picture. The rest of the die is soft. It was put on a charcoal fire and covered up, with the blast shut off. When at a low red heat it was dipped face down in cold salt water and held there for a very short time; then dropped into oil and left there until it was cold. The punch was treated in the same way. It was put in water up to its shoulder. The die and punch were both taken out of the oil and put in the punch press, and up to the time of photographing had been

Fig. 18. 7-16" Punch and Die.

at work nearly two years. Now, if these tools had been heated to a very high degree of heat and cooled off in water until cold, and then drawn back to a certain color, I do not think they would have given the satisfaction they have. I will admit it takes a little practice to become expert at the method of hardening described, but it can be accomplished with a little patience and care and once you make a success of it you will not want to use any other method.

Punches and dies, however, can be more successfully treated by drawing back the temper than can many classes of tools. For instance, a quick and sure way to treat small punches and dies, where quite a few are to be hardened at once, is to pack them in charcoal and leather, equal parts, in a box or pipe. Put in the furnace and leave there until the box is a good bright red; then dump in cold water. Finally clean the pieces and they will be ready to have the temper drawn.

To draw back the punches, have a small sheet iron pan filled with sand, place it over the fire in the forge, and put the punches in the sand—this will result in the temper being drawn uniformly. The punches should be drawn to No. 4 on the temper chart.

To draw back the dies they should be placed endwise on a hot plate of steel and left harder at the top than at the bottom—about No. 5 at the top and No. 4 at the bottom.

Threading Die.—The No. 12 threading die shown in Fig. 19 has been in constant use for over two years up to time of photographing. It has threaded thousands of pieces of one-inch round cast iron parts, and like the other tools mentioned, is as good as new, and was never drawn to a color. It was heated in charcoal to heat indicated by No. 4 on the heat chart. A machine steel cap (Fig. 20) was placed on each side of the die and then all three pieces held together in tongs, shown in Fig. 21, and dipped in the salt water just long enough to harden the teeth of the die, which were exposed to the water through the holes *A,* in the center of the caps. The die was then dropped in oil and allowed to remain until cold. It was then cleaned up and after being sawed apart on one side, as always done with dies of this type, was put to work.

The machine steel caps keep the die soft with the exception of the teeth, which are exposed through the opening marked *A*. If the whole die had been hardened it would have been a difficult tool to draw back and make as good as this one is.

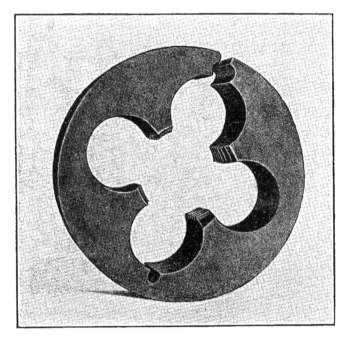

Fig. 19. No. 12 Threading Die.

Treatment of a Broach.—The 1⅛-inch broach, Fig. 22, has 204 teeth or separate cutting parts, and it is a very expensive tool to make. It was dipped at a low heat in salt water, and while the body of tool was still a little red it was thrown in oil and allowed to remain there until it was cold. I have watched this tool at work and it has a hard task to perform broaching holes in machine steel forgings used in making wrenches. Holes are drilled in ¾-inch machine steel, hand and drop forgings. The holes are just large

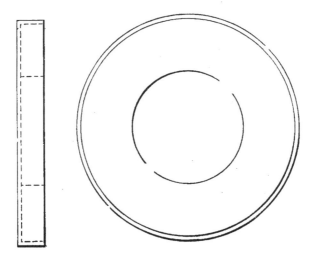

Fig. 20. Cap used in Hardening Threading Die.

Fig. 21. Tongs used for Holding Threading Die.

enough for the small end of the broach to enter. The broach is then pressed down through with a large arbor press. It has been used by men who never worked in a shop before. I am certain from what I have seen that if the broach had not been soft (all but the teeth or cutting parts) they would have broken one on every wrench. This broach, when photographed, had been in use for two years and it was as good as new.

Shear Blades.—In Fig. 23 is shown a pair of shears for cutting off round steel in sizes from ⅛ inch to ½ inch. They have cut drill rod, tool steel, machine steel, cold rolled steel, and any pieces that came along for two years up to the time of photographing, and were still in good condition. They were hardened similar to the punch and die mentioned. Also in Fig. 24 is a pair of long shears, 15 inches long, used for shearing heavy angle iron. These shears were in use the same length of time as the other tools mentioned, and they look none the worse for having sheared thousands of pieces of angle iron from ¼ inch

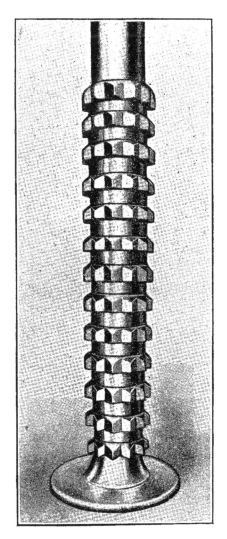

Fig. 22. 1 1-8" Broach with 204 Teeth.

to ⅜ inch thick. · They are quite soft the entire length up to within half an inch of the cutting edge. The shears were dipped at a low heat in salt water, and

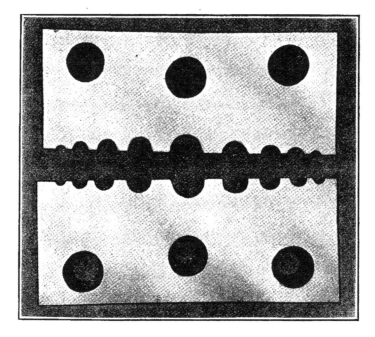

Fig. 23. Shears for Cutting Round Stock.

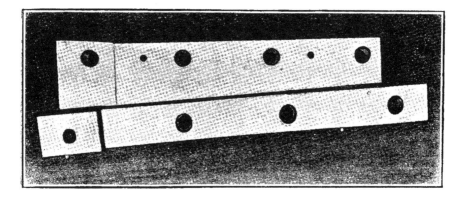

Fig. 24. 15" Shear Blades.

while all but the cutting part or face of shears was still red they were put into oil and remained there until they were cold. They came out straight and hard enough to give the best satisfaction of any tools of their kind I have ever seen in all my shop experience.

Hardening Large Rolls.—In Fig. 25 is shown how to arrange a steel roll that must be treated so as to be very hard on the outside, and that needs to be hardened on the outside only. The roll A is placed upon a center stud between two washers BB. C is a shoulder on the stud against which the washers and the roll are clamped

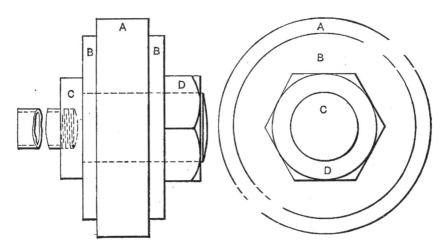

Fig. 25. Large Roll Arranged for Hardening.

by the nut D. The diameter of the roll should be $\frac{1}{16}$ inch larger than the diameter of the stud, and the washers must fit tightly against the faces of the roll. A pipe can be screwed into the end of the stud to handle the work with.

If the roll is a large one, the best way to harden it is to pack in charcoal and put in a furnace; but if you have no box large enough to pack the roll in, or if you have no furnace, just make a large fire with charcoal in your forge, cover the roll entirely over, and let it heat with the coal. When red, uncover a small part of the roll on top, sprinkle cyanide on it and then cover over with coal again and in about five minutes sprinkle on more cyanide. The cyanide melts when it touches the hot roll and will run all the way around the roll. So it is not necessary to uncover the whole roll in applying the cyanide. Care should be taken not to heat the roll to the point where it will begin to scale—a universal rule in the treatment of tool steel.

After the cyanide has been applied the second time, take out the roll and dip in cold salt water. Nothing smaller than a 40-gallon barrel should be used, if you have no regular tank. Do not cool off the roll in the water, but remove before it is cold and put in a tank of oil where it should remain until entirely cold. If a water hose is at hand, it would be well to use it to help chill the roll, but have plenty of salt in the water.

Tempering Springs.—No. 2 on heat chart is about the right heat at which to dip springs. They should be dipped in fish oil until cold and the oil burned off; then dipped in oil again, taken out and laid down to cool off. If you get the right heat, this is all you have to do to be a good spring temperer. If I thought that explaining here how spring steel is made and how it differs from tool steel would help out in tempering, I would do so; but I fail to see how it would, and the same can be said about heating to so many degrees, Fahrenheit. There are books giving this information which are very useful in their place, but if a spring is brought in for the blacksmith to temper, and he does as I have stated, he will have a position longer than he will by starting out to find the percentage of carbon in the steel, and other points that properly concern the scientist.

Drill Jig Bushings.—Heat to a temperature corresponding to No. 8 of the heat chart. Dip in salt water and take out and throw in the oil tank. If the bushing is very light and thin in body, you will want to dip in water and take out quickly and throw in the oil, hot. If it is a large bushing, let the red all disappear before throwing in the oil. I have hardened thousands of drill jig bushings and have never cracked one or known of one being cut by the drill; and yet I have seen tool men harden bushings for drill jigs, clean them up, polish, and draw back to a nice color, and look wise, although they were continually getting the bushings too soft. Next time they would heat them a little hotter and finally crack a lot of them. They would then blame the steel. Whether they did not know or were careless I cannot say; but I *do* know it was an expensive, time-wasting practice.

Tempering a Hammer Head.—In tempering a hammer you will notice that the outside of the face is generally hard and will chip off, while the center is soft. You can overcome this by tempering as shown in Fig. 26. Pour a small stream of water directly on the center of the face of the hammer, and the whole face will be of uniform temper. An old teapot will answer nicely for the purpose.

Tempering Fine Steel Points.—If tempering a pair of divider points, or any small sharp tool, heat to a low heat and cool off in a cake of common brown or yellow soap.

Shrinkage and Expansion.—The shrinkage and expansion of dies, rings, etc., that occurs when dipping has been the cause of a great deal of trouble and wonderment in nearly all shops and I never knew any one who could explain why one ring would shrink while another ring of the same material would expand. In

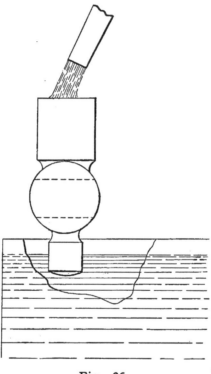

Fig. 26.

my experience the trouble is caused by one side of the ring chilling sooner than the other side. This may seem like too simple an explanation, but try it out and see for yourself.

To prevent expansion in a die or ring when dipping, have some asbestos paste on hand and when the die or ring is at the right heat to dip, cover the inside of the ring with the paste. This will make the ring contract and become smaller. If, on the other hand, you put paste on the outside of the ring, it will expand.

Annealing.—When annealing a piece of tool steel it should not be heated to a temperature higher than indicated by No. 2 on the heat chart. It should be heated slowly, and if in a hurry for the work hold the tool in a dark place and note when the red color has left it entirely; then quench in clear water. If not in a hurry, place the steel on a dry spot and cover with a box or anything that will not burn, but will keep out the air, until the steel is cooled, when it will be the color of No. 1 on the temper chart, with a slight scale which will rub off with the hands. This is about the best and quickest way to anneal.

Another way is to heat the steel to No. 2 and put it in lime. Still another way is to pack the steel in charcoal, heat to No. 3 and let the piece cool off in charcoal. Never put tool steel in a fire of any kind and let it remain over night, as that is the worst treatment tool steel can have. For instance, if a chilled casting is placed in a furnace and removed as soon as it is red hot, and allowed to cool, it will still retain its chilled form. But let it remain in the furnace several hours or all night, and all trace of the chill will be gone when it is cool and the iron will be nothing more than common gray cast iron. If this treatment affects cast iron in the way I have just explained, what can we expect will be the result if we treat fine cast steel the same way? It is foolish to think that the higher a piece of tool steel is heated the softer it will be when cooled off, even though it is cooled slowly. A rolled or hammered bar is always softer than the bar it was made from. This ought to prove what I claim in regard to annealing. If you are not careful in annealing, there is no use being careful about tempering; for, if you spoil a piece of steel in annealing you cannot expect a good tool from it, no matter how you temper it. To get the best results, heat slowly and uniformly in a charcoal fire to color No. 2 and all the way through, and get it

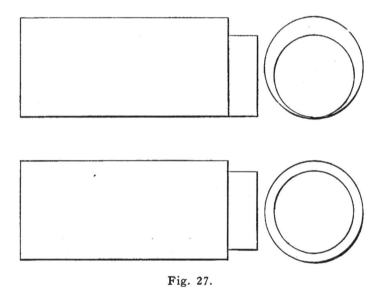

Fig. 27.

out of the fire as soon as you can. (Do not think I am advertising charcoal as I think there is nothing to hinder any one from making his own charcoal.)

If a very thin piece is to be annealed heat to No. 2 and place between two pieces of dry pine board and put a weight on top. When the piece is taken out it will be very soft and also straight.

If you have a tool that will not harden on all sides alike, but will harden only about two thirds of the way around, you will probably find that all the annealed portion was not removed from the bar that the tool was made from. A round mill $2\frac{7}{8}$ inches in diameter should be made from $3\frac{1}{8}$-inch round steel and in all similar cases there should be $\frac{1}{8}$ inch of metal taken from the surface of the bar if the tool is going to be hardened. Sometimes the bar will be carelessly centered as in Fig. 27, leaving the annealed portion on side of the bar. The lower view shows the correct way to center a bar to make a tool for hardening.

General Directions for Hardening.—In bringing this part of the chapter to a close I wish to state that you should never harden a tool of any kind in clear water. Always put in plenty of salt and see that it is well dissolved. To convince yourself, take two tools made from the same bar and harden at the same heat, one in clear water and the other in salt water and note the result. The tool hardened in the clear water will generally crack and pieces will fly off of the cutting edges, while the tool hardened in the salt water will not crack, even if left in the water until it is perfectly cold. Neither should you put oil on top of water and attempt to harden a tool by putting it through the oil into the water. If the reverse of this were possible and we could have the cold salt water on top of the oil, and then dip the tool just hot enough to harden for the class of work required of it down through the cold salt water into the oil, this would be a grand success; but such being impossible, we must do the next best thing, that is, remove the tool from the water and put into the separate tank of fish oil to cool.

A piece to be hardened should be heated slowly and uniformly in a charcoal fire to the correct color for hardening, taking care not to overheat the piece, as that injures the grain of the steel,—injuries that cannot be repaired by drawing back the temper. The piece should then be dipped in salt water, but held there only long enough to harden the cutting edges, after which the tool should be allowed to cool in oil. The oil permits the body of the tool to cool slowly, so that it will be tough and comparatively soft when cold, and at the same time keeps the cutting edges cool so that their tempers will not be destroyed during the cooling process. In the case of hollow tools washers or caps should be used, as already described, to prevent the inner surfaces from hardening when dipped in the salt water to harden the cutting edges. Tools heated in this way not only give the best of service, but will not warp and crack.

CHAPTER IV.

HIGH-SPEED STEEL.

There are many different makes of high-speed steel, each having a different name and each one is claimed by its maker to be a little better than the others. I have had experience with nearly all kinds and makes since they were introduced and will try to explain as clearly as possible how to handle these steels.

The peculiarities of high-speed steel are that it must be heated to a very much higher temperature than the ordinary carbon steels in the process of hardening and that it withstands a much higher temperature when cutting metal and consequently allows a greatly increased cutting speed. Certain chemical elements are introduced in these steels which unite with the carbon in the steel and produce carbides of great hardness and durability at high temperatures.

The first steels of this character were the self-hardening varieties, such as Mushet steel, but within a few years have come the air-hardening and oil-hardening high-speed steels, which are capable of a very much greater output of work than any steel previously made.

To Distinguish High-speed Steel from ordinary carbon or tool steel, hold a piece of the steel on an emery wheel and note the color of the sparks. High-speed steel produces red sparks, while tool steel and machine steel produce white sparks.

Cutting Off High-Speed Steel.—High-speed and self-hardening steels should be heated to a cherry red if not annealed and nicked around as at *A* in No. 3, Fig. 1, and broken while hot. If the bars are not too large the smith can nick them on the hardy and by using a tool such as *B,* Fig. 1, can easily break the pieces off. The bar to be broken should be so placed in the tool that the nicked part will rest on edge *C,*

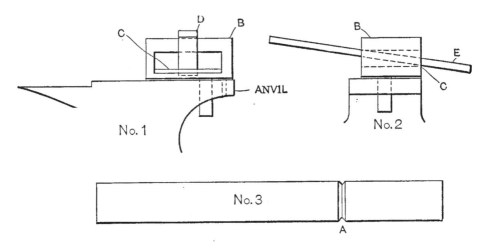

Fig. 1. Cutting off Steel.

which should be tempered. By holding up tight on end *D,* and striking a blow at *E* the piece will fall off. Small bars of high speed steel are sometimes nicked around on the edge of an emery wheel preparatory to breaking. This is a clumsy method, but does well in some cases.

Small stock, such as $\frac{1}{4}$, $\frac{5}{16}$, and $\frac{3}{8}$ inch square, which is used in tool holders, can be nicked and broken cold, but the smith should use a file to test the steel that is brought to him to make sure that it is annealed before he attempts to nick it with the hardy or cold chisel. Air-hardening and Mushet steels are seldom if ever annealed when purchased and if one tries to nick them cold, before annealing, pieces are liable to fly from the chisel or hardy and injure someone— most likely the smith himself.

By the use of the tool shown in Fig. 1, pieces can be broken off from small bars as rapidly as the smith can handle the bar. In shops where these pieces are used in tool holders it is a good plan to take a bar, or even two or three bars of annealed high-speed steel, mark off the desired lengths, nick around as stated, and break. The smith can not only do a faster job, but a better job of hardening where there are a hundred or so pieces of steel to harden than where he is bothered with one piece at a time at half-hour intervals.

Treating Self-hardening Steel.—Self-hardening steel should be worked or forged at a cherry heat and it has to be clipped to shape more than any other steel, as it is very brittle, even when hot. After forging and grinding to shape it should be reheated to a cherry red and let cool in the open air.

Forging High-Speed Steel.—This steel should be heated slowly and carefully, and kept well covered in a fire with a good solid bottom, blacksmithing coke being used for fuel. It should be worked at a very high heat, with heavy, rapid blows, and should never be hammered when below a cherry red. When it reaches this point it should be reheated. Always accomplish as much as possible at each heat.

A tool of any desired shape can be forged from air-hardening steel if properly heated and worked and the same applies to oil-hardening steel. In fact, all high-speed steels except self-hardening or Mushet should be heated and forged in the same manner. After forging to shape the tool should cool off slowly and then it is ready for hardening.

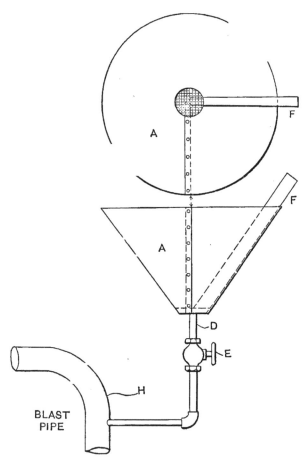

Fig. 2. Device for Air Hardening.

Device for Air Hardening.—Compressed air under the pressure usually carried is not a good thing to use for air-hardening tools. A milder blast, well distributed over the tools is more satisfactory and nothing that the writer has seen can equal the following arrangement for air hardening:

Tap the pipe that furnishes the blast to the forge

for a ¾-inch gas pipe and connect with this a funnel made from light sheet iron as shown in Fig. 2, where A is the funnel, D is the connecting pipe to funnel and E is a valve to shut off the blast when no hardening is being done. The funnel should be about 12 inches diameter at the top and about $1\frac{1}{4}$ inches at the bottom and the bottom should be covered with a fine mesh sieve to prevent small tools from dropping into the supply pipe. Tools to be hardened are placed in the funnel as shown at F, with the cutting point at about the center of the sieve.

Heating for Hardening.—In reheating a tool for hardening, after it has been forged and allowed to cool off, it should be treated on the point only, and in a charcoal fire. It should be heated about as hot as a charcoal fire will make it, or until the point begins to flow; but it is not necessary to burn the point and blister it as some do. The latter is liable to happen if blacksmithing coal is used instead of charcoal. If the tool is heated back very far from the point, it is liable to break when tightened in the tool post.

In heating the tool it should be protected from the blast from below and the air above and should be reheated slowly at first to make sure that the steel is brought to a uniform temperature throughout. Then put on a little more blast and bring to a flowing heat quickly. Do not take the tool out of the fire and look at it and replace it a number of times. This is bad practice.

When the desired heat is obtained, quickly remove from the fire and place in the funnel with the air supply opened, as already explained.

If the steel is overheated the air will cause the steel to burn instead of cooling it off. For your own satisfaction disconnect the valve from the pipe below the funnel, so as to get a direct air blast, and heat a piece of ¾-inch round high-speed steel about four inches from the end, until it is very hot,—dazzling white heat. Place the steel in the blast near the end of the pipe and note the result. If the steel is hot enough the air will cause it to burn apart. You will see from this what injury may be done to a lathe or planer tool, or any high-speed tool steel, if heated too hot.

Hardening with Cyanide.—Many high-speed steels are treated as follows: Heat the point to a yellow and dip in cyanide; then replace in fire and heat very hot and drop in a tank of kerosene oil. Sometimes, also, the steel is hardened by heating very hot and dropping in oil, without using the cyanide. All manufacturers furnish directions for treating their special brands of steel, which will indicate to the smith what special methods will need to be followed in any particular case.

Hardening Milling Cutters and other Expensive Tools.—For heating high-speed milling cutters and other similar tools having delicate cutting edges, about the best arrangement is a furnace built on the order of the one in Fig. 3, in which the two lower views are the front and side elevation and the upper views show details. To construct the furnace, make a foundation A from cast iron of the required size. Then take two pieces of boiler plate ½ inch thick and bend like B and rivet to the top of the foundation. The tuyere E is made of cast iron or light boiler plate. There

HIGH-SPEED STEEL 113

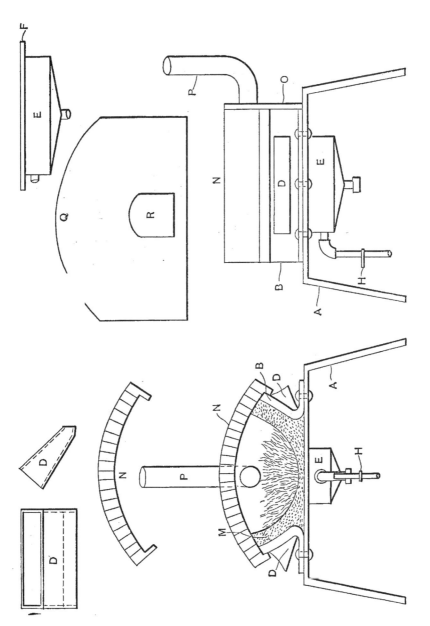

Fig. 3. Furnace for High-speed Steel.

should be an oblong hole through the top of the foundation to let the tuyere through and allow it to rest on the flanges F. The tuyere must have a hole in one end for the blast and an outlet through a nipple at the bottom for the cleanout. A cap screwed on the end of the nipple closes the cleanout when not in use. By removing the cap and opening H in blast pipe the dirt will be blown out onto the floor. The tuyere should be drilled with $\frac{3}{8}$-inch holes on top for the blast.

Now the most important feature of this forge is the two funnels D extending through the sides of the furnace to the outside edges of the tuyere. They should be of the same length as the tuyere, about eight inches wide at the top and four inches at the bottom. Their purpose is to feed fuel to the bottom of the fire when heating work to be hardened without disturbing the fire and causing air to reach the tool being heated. After getting the funnels in place, the furnace must be lined about four inches deep with Portland cement and sand, mixed in equal parts, leaving the lower ends of the funnels open. In the illustration M shows the cement lining.

A light steel frame should be made to hold the firebrick cover N together. This cover can be removed at any time desired. A back O is made from boiler plate with hole for stack P to connect with the chimney or down draught, as the case may be.

A door Q should be provided for the front with opening R large enough to allow large tools to be passed through.

After building the fire in the furnace cover the tool over with coke or wood charcoal, replace the door, and supply the rest of the fuel through the funnels on the side, which feed the bottom of the fire. With this arrangement you will not have to turn the tool very much and it will heat uniformly. Do not use hard coke, as it will injure the fine edges of the cutters. If you use charcoal, put on a good lively blast, as you cannot overheat the tool. Harden by dipping in a tank of fish oil. If there is much of this work to be done the oil should be kept cool by setting the oil tank in a tank of running water.

If thin milling cutters, $\frac{1}{2}$ inch thick or less and four or five inches in diameter, have been made from high-speed steel and are to be hardened, it would be advisable to use the stud arrangement shown in Fig. 14, Chapter III. Put on a good blast, turn cutters constantly and when hot enough dip in fish oil. I have hardened cutters of this kind in this way and they did 20 per cent more work without grinding than the same kind of tools that came already hardened from the makers of the steel. With a little practice the smith can become expert in handling high-speed steel; in fact it is not so easily ruined as high-carbon tool steel.

Hardening long Blades.—Fig. 4 shows how long, thin, narrow, high-speed steel blades may be hardened successfully without cracking or warping. (You can harden them to a certain extent in oil, one at a time, but not hard enough to be satisfactory.) Make three clamps, as shown in Fig. 4. Put one on each end and one in the center of the blades as shown in Nos.

1, 2, and 3. The clamps can be made wide enough to hold ten or more blades, which should fit tight in the bottom part C. The top part of clamp A should bind on edge of the blades and be tightened with cap screw B. Pack in fine wood charcoal, using a piece of gas

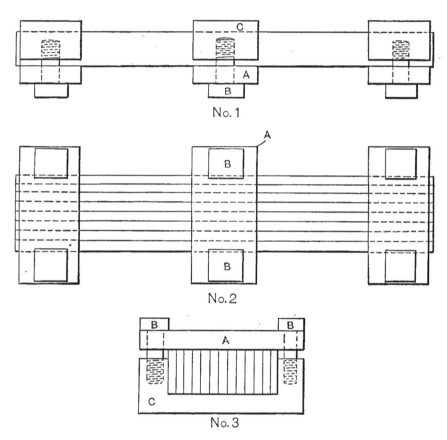

Fig. 4. Device used in Hardening Blades of High-speed Steel.

pipe large enough for the purpose, and plug both ends with asbestos cement. Then place in the oven and heat to a very high heat, almost lemon. They should remain in the oven two hours, then take out and dip endwise in hot salt water that you can just hold your hand in; dissolve one pint of salt to a pail of water.

Keep the blades in the water until as cool as warm water will make them, then remove to oil and leave until cold. You will find on taking them out they will be very hard and also straight. Water drawn from the boiler is very good for this purpose. It is better than water drawn cold and heated with a hot iron as usually is done in blacksmith shops.

Annealing High-speed Steel.—One of the most successful methods for annealing high-speed steel without lessening its hardening qualities afterward is to pack the tools to be annealed in fine charcoal and heat to a cherry red in the furnace and keep in the furnace for two hours after the box becomes red. Then remove the box from the furnace and place in a dry place to cool off. The cover should be sealed air tight, as the tools must be kept from the air while hot. You can also put high-speed steel tools in the forge and cover with charcoal, leaving over night. This will soften the steel and put it in shape for machining. A furnace suitable for annealing is described in the next chapter.

CHAPTER V.

CASEHARDENING AND COLORING.

Processes in which a Furnace is Required.

While the charcoal forge affords a convenient and an effective means for heating work of the character already described, a furnace in which a uniform heat can be maintained through a long period of time will be found necessary for hardening some tool steel parts and for casehardening and coloring, where much of such work is to be done.

The Furnace.—In Fig. 1 is a view of such a furnace, designed to burn hard coal, which is economical to build and operate and gives very satisfactory results. In the appendix are sheets containing complete working drawings of the furnace. Many small concerns could build a furnace like this and harden all the machine steel parts necessary in the construction of their machinery and tools. The furnace can be built from common brick and firebrick, without the use of any large tile, and is adapted for mottling and coloring, as well as casehardening and pack hardening. In the front view of the furnace in the appendix, the left-hand door is for firing and the right-hand door for placing and removing work. Below are the ashpit and flue openings, covered with sheet iron caps, for cleaning out. The blast pipe (No. 5) is at the left.

The second view is a cross section. Here A is the firebox, B the bridge wall, C the bottom of the oven on which the work is placed, and $E\ E$ and F flues for carrying off the waste gases. These flues are so arranged that the gases circulate all around under the bottom of the oven before passing to the bottom flue F which carries them to the stack. The gases have to pass through a single opening in order to reach the lower flue, and this opening is controlled by a damper.

Fig. 1. Furnace for Casehardening, Pack Hardening, etc.

The longitudinal section, Fig. 3 in the appendix, shows the flues and damper, and their arrangement, very clearly.

The last view, which is a sectional plan, further shows the arrangement of the firebox and oven. It will be noted that there are four square openings, one at each corner of the oven, through which the gases pass down to the flues underneath; while the passage from these latter flues to the flue leading to the stack is at the center, controlled by the damper, thus giving the best possible heating effect.

The following is a numbered list of the most essential parts of the furnace:

1—Cast-iron buck stays.
2—⅝-inch stay bolts.
3—Door frame 1¾ inches by ¾ inch iron.
4—Sheet iron caps for flues.
5—Blast pipe, 1½-inch gas pipe, slotted.
6—Damper.
7—Damper support.
8—Cast-iron grate.
9—Grate support.
10—Blast shut off.
11—Stack.

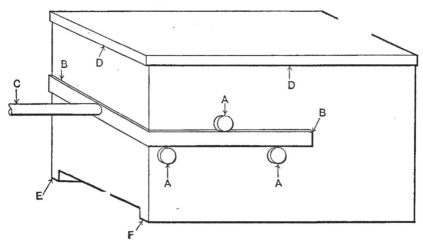

Fig. 2. Box for Packing Pieces for Casehardening.

Packing for Casehardening.—In packing machine steel pieces for casehardening in a furnace, we should

CASEHARDENING AND COLORING

have boxes of various sizes and also an assortment of gas pipe of different lengths and sizes. The pipe should have a plug riveted in one end, the other end to be plugged with asbestos cement after the pieces have been placed in the pipe. Boxes of cast iron are the best to use, as boiler plate is expensive and gets out of shape quicker than cast iron. The cast iron box becomes malleable after using it a few times.

Fig. 2 shows a fair sample of a box for ordinary casehardening. It should have three bosses A on each side and fork B should be made to pass through between the bosses and fit to the side of the box. The fork should have a handle C made from round steel, with a hand hold on the end. There should be different sized forks for different sized boxes, and the boxes should have a web E at the bottom, the entire length on each side. This allows heat to pass under the box, whereby a more uniform heat can be obtained. After the parts to be hardened are packed in the box the lid should be put on and sealed along edges D, with asbestos cement moistened to make a paste.

Directions for Casehardening.—Raw bone is generally used in hardening machine steel parts. There are different grades of granulated raw bone, Nos. 1, 2, 3 and 4. No. 1 is the best. No. 4 is a grade generally purchased for poultry food and is not so well adapted for casehardening. One part raw bone and five parts fine wood charcoal makes an excellent mixture for casehardening. Raw bone can be dried and used the second time by mixing new bone with it, about equal parts. Bone black is also used and gives good results.

The depth of carbonizing is determined by the length

of time the boxes are left in the furnace, the time ranging from three to 12 hours.

If a very fine grain is desired, the boxes should be left in the furnace for about six hours after becoming a very bright red, and then set out to cool without removing the lid, as by this means the parts will continue to carbonize until nearly cold. When cold, the parts to be hardened should be placed in the box without any mixture and reheated to a bright red and dumped into cold water. This second heating without any packing increases the depth of the hardening.

Sometimes work is required to be soft on some parts and hard on others, as would be the case with a hub sprocket. This result can be obtained by packing fine, dry sand around the parts to be left soft. If the pieces are so shaped as to make it impossible to pack dry sand around the parts to be left soft, make a paste from ordinary clay and cover the parts enough to protect them from the carbonizing mixture.

Tank for Casehardening Work.—The best kind of a tank for receiving work from the casehardening furnace is shown in Fig. 3. In taking the boxes from the furnace set them on the steel plate A, supported by the braces E. Remove the cover from the box and dump contents on sieve B. The bone, or whatever the work is packed in, falls through the sieve to the bottom of tank X, while the parts to be hardened roll into the water at C. Another sieve is suspended in the water by small steel rods. After the work is all dumped this inner sieve can be drawn up and the hardened pieces taken off, clean, ready to be delivered to the department they are intended for. This is much better than dump-

CASEHARDENING AND COLORING 123

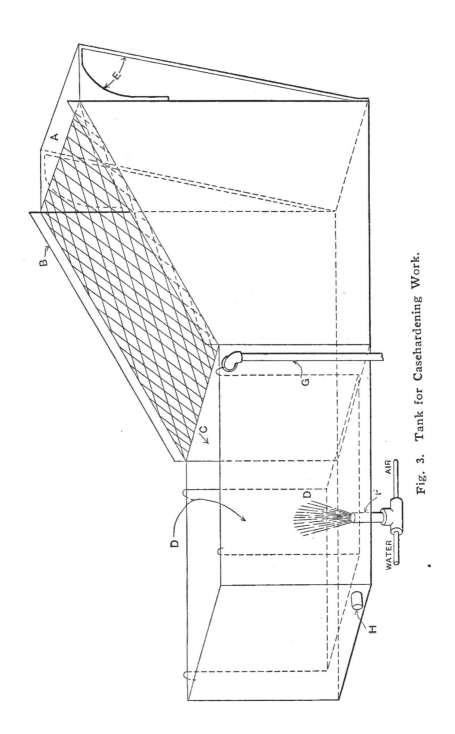

Fig. 3. Tank for Casehardening Work.

ing the whole contents of the box into one opening and having to sort out the parts afterwards. It is also better because dumping the contents together into the water tank retards the hardening, since the water cannot then reach the work directly.

Several sieves can be provided, with different meshes; the coarse ones for heavy work and the finer ones for small work, which avoids letting the small work drop through the sieve into the part of the tank intended for the bone only. Another good feature is that the bone can be taken from tank X and dried, to be used again for mottling and coloring work, as described elsewhere in the book.

In Fig. 3 F is the inlet for water; G the overflow; H the outlet for cleaning the tank, which should be done once a week. There should be a steady flow of water at all times when dumping. The air connection to the supply pipe is for use in mottling and coloring as will be explained later.

A tank of this kind is shown in Fig. 1 in connection with the furnace there illustrated. This was made of pine plank covered with sheet iron inside and outside, and let into the ground about 12 inches, having 24 inches above the ground, thus making a large body of water to dump the work into. Light boiler plate could also be used for making the tank.

Tools from Machine Steel.—Bending and forming dies for punch press work, if made from machine steel, can be carbonized and hardened and will give better satisfaction than tool steel because they are hard only on the outside and not so liable to break.

Another advantage in using machine steel is because it is cheaper and is much easier to machine in the tool room.

Pack Hardening.—Casehardening is used for machine steel parts, while a similar process for tool steel parts is called pack hardening. In Chapter III. reference was made to this process in connection with the subject of reamers. Heating tools for hardening in gas furnaces, or any open furnace, without packing, oxidizes and decarbonizes the steel, thus making it almost impossible to obtain a uniform hardness. Lead baths have their disadvantages, as already explained. By packing in fine charcoal, however, heating in the furnace and dipping in salt water, in accordance with the instructions given, satisfactory results will be secured. If you pack in raw bone, you will get too much carbon in the cutting parts, thereby making the tool brittle. Charcoal alone for heating and salt water for dipping has been worked out to the writer's satisfaction.

Pack Hardening Long Pieces.—Pieces that are too long to heat in a charcoal forge, shown elsewhere in the book, should be packed in fine charcoal, using a box similar to that shown in No. 2, Fig. 4. It should have a cover, and after the work to be hardened is packed in the box, the cover should be sealed. The box should be long enough to let the pieces clear about $1\frac{1}{2}$ inches at each end. If using a furnace and firing with hard coal, and if the furnace is at normal heat inside, place the box in the furnace and leave for about one hour, or a little more; perhaps $1\frac{1}{2}$ hours.

When taking from the box to dip, use two pairs of tongs, if the piece is 24 inches long or over. Have a pair of tongs in each hand, grasp about six inches from each end, and when putting the piece in cold salt water be sure to put it in quickly and let it go straight down, then come up slowly. If the red has disappeared, remove to the oil tank to draw back, and you will have a tool that will give satisfaction.

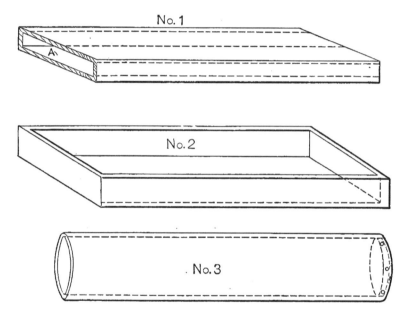

Fig. 4. Boxes for Pack Hardening.

Box No. 1 in Fig. 4 is another type suitable for pack hardening. With this, however, it is necessary to shove the work to be hardened in at end *A*. When the pieces are ready to take out, you have to take hold of the end and draw the entire piece out of the box. Before getting hold with another pair of tongs, this operation is almost certain to bend the piece, causing it to warp when put in the water. If you have no box like No. 2, use a piece of pipe as shown in No. 3, Fig. 4. Rivet a plug in one end, and after the parts to be hardened are packed in, plug the other end with asbestos cement.

A pipe has the advantage over any box for heating tool steel for hardening, because if it is not heating evenly you can roll the pipe bottom side up, thus obtaining the desired result; but it is no better than box No. 1 for getting long pieces out of, and for this reason alone box No. 2 is better than either of the others.

In packing tool steel pieces to be hardened, use nothing but fine wood charcoal. If any raw bone is used it will render the parts brittle. Try the charcoal alone and note results and the amount of hard work the pieces will stand.

Testing Work.—A good way for a beginner to know when the work in boxes is hot enough is to put a few ¼-inch rods through holes in boxes or pipe, and draw one out after the box has become red; if the rod is dark, return it to the box, so as to prevent the carbon escaping, and after a while draw another rod—not the one drawn first. If the second one drawn is red, cool quickly and break. If this is tried out by the workman on this class of work he will soon become expert enough to gauge his heat without the rods. This method is advisable both in casehardening and in pack hardening, more so in the latter, as in the former process the work is generally left in the furnace long enough to insure a good, uniform heat.

Pack Hardening Thin Cutters.—In a shop where a quantity of cutters is to be hardened, they should be arranged as in Fig. 5. This shows ten milling cutters on one stud, with a washer on each end, small enough in diameter to allow the cutters to harden well below the bottoms of the teeth, and at the same time large

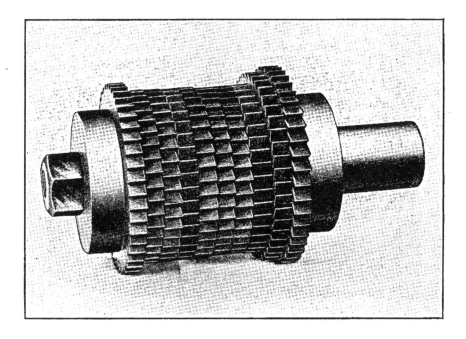

Fig. 5. Cutters Treated by Pack Hardening.

enough to prevent warping or cracking. The cutters shown vary in diameter from three inches to $4\frac{1}{2}$ inches and have $\frac{1}{2}$-inch cutting face. Thin cutters of $\frac{1}{16}$-inch face and upwards, should be hardened in this manner, as it does away with any possible warping and gives a uniform hardness.

Fig. 6 shows the stud, washers and nut before the cutters are put on for packing. Just enough of the stud should be left at the large end to take hold of with the tongs. The cutters should be put on and the nut tightened as tight as possible. Then pack in a pipe large enough to allow ½-inch clearance between the teeth and inside of pipe, pack in fine wood charcoal, seal the ends of pipe with asbestos cement and heat in the furnace. The time required in the furnace is from one to two hours. If the furnace is at normal heat when the work is put in one hour will do. If no furnace is at hand, then place the pipe over a large forge and cover with coal or hard coke. If the pipe is not

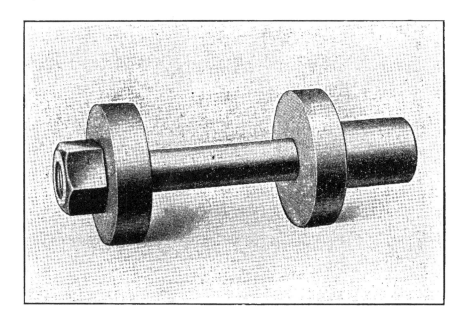

Fig. 6. Arbor used with Cutters.

heating uniformly, roll it over. This is where the pipe has the advantage over the box for pack hardening. The cutters shown here were dipped at a bright red in cold salt water and while still quite warm were removed to a tank of fish oil to draw back. They came out straight and were hard from the outside cutting edge to within ⅛ inch above the washers. This is the best possible way to do a first-class job on work of this kind when produced in quantities. Of course, if only one or two are to be hardened the method explained elsewhere in the book is the best.

Cyanide Hardening.—There are many different methods of hardening with cyanide potassium. Sometimes, and with some classes of work, we heat the parts to be hardened in the forge to a bright red, sprinkle with dry, powdered cyanide and cool in cold salt water; and for another class of work we use a crucible. The parts to be hardened are suspended by wires in the crucible of cyanide long enough to attain the same heat as cyanide, then dipped in cold water. *Cyanide being very poisonous, care should be used, both in handling and in working over it. Do not breathe the fumes.*

A good method of hardening with cyanide is to put the parts to be hardened in a box used in bone hardening, put on cover and heat in furnace. When work is cherry red, sprinkle with cyanide and return the cover. In ten or fifteen minutes repeat the sprinkling, then dump the contents in cold water. This gives a good hard surface.

In heating work in the forge for cyanide hardening it should not be scaled, as the cyanide will be coated on the outside scale, and when the piece is put in water the scale will drop off, leaving the work soft. It should be heated about the same as when hardening tool steel, and never hot enough to scale.

About the most successful method of cyanide hardening developed by the writer is as follows: Put a good quantity of cyanide in one of the casehardening boxes, put the lid on the box, place in the furnace and heat to a bright cherry red. At the same time place the parts to be hardened in an open furnace and when at the same heat as the cyanide remove lid from the box and put the parts into the cyanide. Leave in about fifteen minutes, and then take box from the furnace. Take the parts out, one at a time, and cool off in a tank of cold kerosene oil. This hardens very deep and also toughens the steel. For crank shafts and such pieces this method cannot be equalled.

Mottling and Coloring.—The writer has experimented in this line of work and has obtained some of the best results on record. A sample of the work is shown in the appendix. The following method not only colors but hardens deep enough for the class of pieces desired to be colored, such as wrenches, cranks, levers, bolts, etc. Pieces to be treated must be polished, and no grease of any kind should be on them. Use 10 parts charred bone, 10 parts charred leather, 10 parts wood charcoal, one part cyanide. All should be pulverized and thoroughly mixed together. If the colors are too gaudy, leave out the cyanide, and, if still less coloring is desired, leave out the charcoal.

When packing for this class of work, use a piece of gas pipe, of the required size and length, and plug the end with asbestos cement. Leave the work in the furnace about four hours, heated just to a cherry red. If heated too hot no colors will appear.

If much of this class of work is to be done, a furnace should be supplied for this alone, as a furnace heated for casehardening is too hot for coloring.

When dumping this work, an air pipe connected with the water pipe in tank should come in at the bottom. Turn on the water and enough air to make the tank full of bubbles and quite lively. Then take the pipe from furnace and remove the asbestos plug. Hold the end of pipe close to the water before letting the work appear, as the work should not be exposed to air before striking the water. Dump in the center of the tank where the most air is (see *D,* Fig. 3) and then draw up the sieve and remove the work to a pail of boiling water, drawn from the boiler. Leave in the hot water for about five minutes and remove to a box of dry sawdust. After ten or fifteen minutes take out and remove all dust and apply oil or lacquer.

How to Get the Charred Bone and Leather.—Fill your boxes with raw bone, put cover on and place in furnace, when the furnace is not too hot. For instance, if it is your custom to let your furnace fire out Saturday night and relight Monday morning, then Saturday night would be a good time to char the bone, if you are using a hard coal furnace. If the bone turns a dark brown it is enough, the object being to remove all grease. If any grease remains in the mixture there will be no colors.

The leather may be treated in the same way as the bone, and should be charred so that it can be pulverized quite fine. Leather chars much easier than bone, and any old scrap belting, such as found in all large factories, can be used.

Coloring with Cyanide.—Pieces of steel can be hardened and colored nicely by heating cyanide to cherry red in a crucible or box placed in the casehardening furnace. Place parts to be hardened in the red hot cyanide and let remain twenty minutes, then pick out one at a time and drop them in the tank with air bubbling freely up from bottom.

Coloring by Heat Alone.—Almost any color desired can be obtained on machine or tool steel by using sand in an iron box and heating over a forge. The sand must be quite hot and thoroughly mixed, so that the heat will be uniform.

Place the pieces to be colored in the sand and watch closely, as the color comes quickly and changes from light straw to dark blue very rapidly. When the desired color is obtained dip in water just enough to check and let cool off in sawdust. Steel colored in this way is, of course, soft, and the coloring is of no use, except for its appearance.

CHAPTER VI.

BRAZING—GENERAL BLACKSMITHING.

Brazing.—This class of work requires more care and attention than is generally paid to it. Some will bring to the brazing furnace a lot of rusty, greasy parts, driven together so tightly with a sledge hammer that the spelter cannot enter and when applied will simply spread around on the outside of the work. When the brazing is completed it may look like a good job to the inexperienced eye, but such pieces cannot be brazed satisfactorily and will show up bad after being in use a while.

Parts to be brazed should be clean and free from scale and rust, with enough space in the joints to allow the spelter to sweat in between the pieces and hold the surfaces together.

Brazing Furnace.—In Fig. 1 are two views of a good brazing furnace, made from firebrick. The frame of the furnace consists of a cast iron plate A, with four holes drilled and tapped for legs B made from gas pipe. There are braces C around the legs at the bottom, to prevent spreading. The walls of the furnace are built up of firebrick on top of the flat plate and another iron plate is used to cover the top. This latter plate can be removed at any time to allow the brazing of such pieces as would have to project through the top of the furnace.

This furnace is shown equipped with three burners using gas for fuel. Where gas cannot be had it is common to use gasoline, burners for which can be bought ready for use. An air blast must be provided for either type of burner. If gas is the fuel, burners can be made by using pipe fittings, as in Fig. 1. The front view shows two side burners made from 1¼-inch pipe fittings, gas connections being by pipes *F F* and air connections by the rubber hose *G G*. By using the hose the burners can be changed to slightly differ-

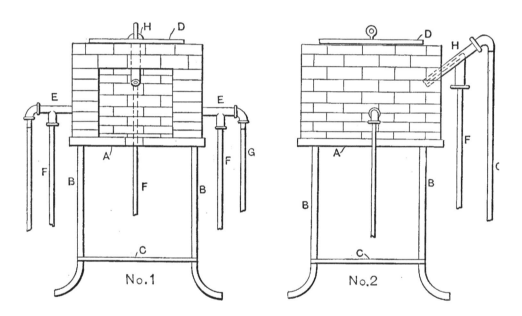

Fig. 1. Brazing Furnace.

ent angles, as required, and by using a 45-degree T a burner like that in the side view at the back of the furnace can be made, which, in this position, blows down directly on the work. This makes a splendid arrangement for brazing most all ordinary pieces. A small ¼-inch gas pipe should be fitted on the inside of the T forming the burner and come within one inch of the end of the T. The construction is shown by the dotted lines. This is connected with the air supply, the gas flowing around the outside of the ¼-inch pipe.

For brazing pieces that cannot be put in the furnace, take one burner off and connect it up with rubber hose to both gas and air connections and in this way the blast can be played on parts that cannot be reached in the furnace. The disconnected burner is shown in Fig. 2. Here A is the air tube, B the 1¼-inch T and

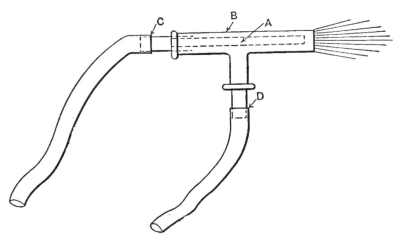

Fig. 2.

C and D the hose connections. All valves should be arranged within easy reach of the man doing the brazing, as it is necessary to regulate them frequently.

Spelter.—Boracic acid and fine brass chips make the best spelter. The pieces of brass should not be larger than common barrel salt. Soft brass wire of different sizes is also good, especially when the part to be brazed is difficult to get at with a spoon. The boracic acid and brass should be mixed, equal parts, in a very shallow pan. If pan is too deep, the brass, being heavy, goes to the bottom, leaving the boracic acid at top and poor brazing is the result. Spelter is used dry, and applied with a spoon with a handle about 36 inches long; and it is not applied until the parts to be brazed have become hot enough to melt it quickly.

Directions for Brazing.—Fig. 3 shows a piece of steel tubing and a machine steel piece to be brazed together. The hole in No. 1 should be $\frac{1}{64}$-inch larger than the swaged end X of No. 2. It should be placed

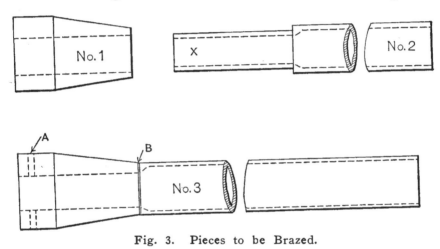

Fig. 3. Pieces to be Brazed.

in position and pinned, as shown at A, No. 3. Before putting together, part X should be ground on an emery wheel to insure a clean, thorough job.

Make a paste from boracic acid and water and put a good coating on part X before fitting the two pieces together. Then pin as stated and put a heavy coating around joint B. Prepare a number in this way and let the acid dry on, put into the furnace and when the piece is hot enough to melt the spelter apply quickly with a long handled spoon, turning the work with one hand. As soon as the spelter flows freely through the joint, remove the work from the furnace, as it injures the work to overheat it. If pieces are to be machined in any way, they should, on removing from the furnace, be put in lime, which leaves them softer than when left to cool off in the air.

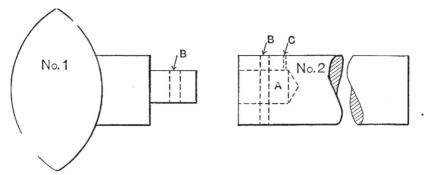

Fig. 4. Pieces to be Brazed.

It is a wrong practice to keep cooling off the spoon in applying the spelter. Let the spoon remain hot and use it in guiding the course of the spelter around the part to be brazed. By cooling the spoon and dipping into the spelter while wet, the blast from the burner will blow the spelter from the spoon before reaching the joint or part to be brazed; but if the spoon is hot the spelter will adhere to it long enough to allow applying to the joint. The part being brazed is so much hotter than the spoon that it melts the spelter, which flows from the spoon freely. The spelter will sweat into the joint three or four inches if the parts are clean and the heat is uniform.

In Fig. 4 is a job of brazing where piece No. 1 is to be brazed into a hole in piece No. 2. Plenty of spelter should be put in the bottom of the hole at *A*. Then put the pieces together and pin (see dotted line at *B*).

A small vent hole C should be drilled in work of this class to prevent an explosion.

Brazing Cast Iron.—Cast-iron pieces, such as broken parts of machinery, can be brazed just as easily as machine steel or wrought iron. The cast iron to be brazed should be free from all grease and coated with oxide of copper. Then pin the parts together and braze the same as machine steel parts are brazed. The oxide of copper decarbonizes the cast iron for a sufficient distance to allow the parts to unite, by the spelter filling the place formerly occupied by the carbon. Care should be taken in heating cast iron as it melts at a much lower heat than most metals.

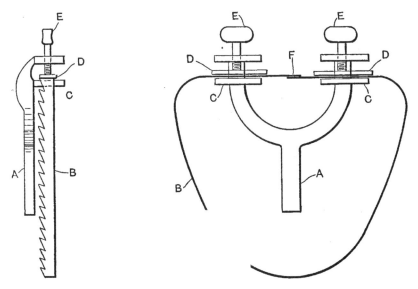

Fig. 5. Brazing a Band Saw.

Brazing a Small Band Saw.—This seems an impossible job to any one who has never seen the operation. Make a double clamp for holding the ends of the saw, as shown in Fig. 5, the clamp being held in the vise or in a swageblock by the part A. Then grind the ends of the saw tapering and bend the saw with the tapered ends overlapping, in position for brazing. Rest the saw on the bottom of the clamp and place a flat piece of steel D on top of the saw and tighten down with the thumbscrews E. Then put some silver solder between the ground ends of the saw and cover with boracic acid paste. Have a pair of tongs with jaws about one inch square that fit close together, heat the tongs to a bright red heat, and close down on saw at F. Hold the tongs in position for about 20 seconds and remove, and the trick is done.

If no silver solder is at hand, use a 10-cent piece. Flatten out quite thin and put a small piece between the ends of the saw. The coin contains copper enough to allow it to do the work just as well as silver solder.

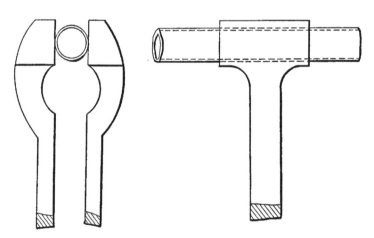

Fig. 6. Pipe Bending.

Bending Gas Pipe for experimental purposes is an operation that troubles many blacksmiths. To do this without flattening the pipe at the bend, heat the pipe to a low red and put in the vise as in Fig. 6. Tighten the vise just enough to prevent the pipe slipping out, putting the hottest part of the pipe at center of the vise jaws. Then with one man at each end of the pipe, bend it downward or upward,—never outward. A pipe can be bent to a perfect radius in this way in a few heats.

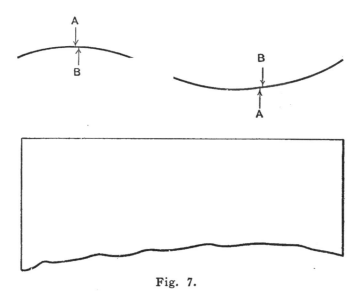

Fig. 7.

To Straighten Thin Sheet Steel after it has been warped by heating and working, some smiths heat it quite hot and put a heavy weight on it, expecting that when the piece is cold it will be flat and true. But not so. The hump will be there just the same. Fig. 7 represents a piece of thin steel that has become bent or warped. Now if you heat the piece and hammer it on the high side *A,* you will make it worse. That side of the metal is longest already and hammering expands it still more. But by striking a few slight blows around the concave side, *B,* and expanding it, you will find it will come back straight. This must be done cold.

The General Blacksmith and Horse Shoer.—While the contents of this book up to the present are of some interest to the horse shoer and general blacksmith, what follows is of interest to them only. Some of the best blacksmiths have commenced in a general jobbing shop. After serving a couple of years in such a shop, an ambitious young man may secure employment in a large shop, and in a short time become a very useful mechanic.

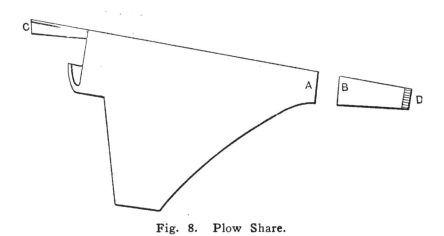

Fig. 8. Plow Share.

Repairing Plowshares.—Fig. 8 shows a plowshare. About twenty-five years ago plowshares were generally forged from wrought-iron plates, and when the wing wore quite narrow it was laid along the share with a piece of steel, and very often had to be pointed, staved up and made square on end. Cast-iron shares soon drove this class of work away from the general smith. Now a cast-iron share becomes useless as soon as the point wears rounding on the bottom and becomes thin like a lance. It no longer takes hold, and is thrown in the scrap heap. The writer has repaired hundreds of cast-iron plow points which have reached this condition. Break off a piece from the end of the point and grind a good, flat, straight surface (see *A*, Fig. 8). Then forge a steel point the proper shape as shown at *B*. Now heat both pieces, using borax, watching the cast iron closely, but get the steel part as hot as possible. When the pieces are hot enough, place the share on a level plate, with a support at the end *C*, then press the steel piece up against the cast iron and they will unite. Do not hammer; let them cool and then reheat point *D*, and dip in cold salt water to harden. It will be found impossible to break a piece joined in this way.

Shoeing to Prevent Interfering.—It used to be the custom among a great many smiths, in order to stop or cure a horse from interfering, to pare the hoof nearly all away at the inside of the foot, and bend the inside of the bar of shoe around under the heel. This makes matters worse, as an interfering horse always wears off the inside of his hoof more than the outside; and again, if you notice a horse that does not interfere or strike, you will observe his hoof is worn down on the outside more than the inside.

By treating the horse that interferes so he will throw the joint out he will never strike. This is done by building up the *inside*. Do not pare it. **P**are the outside and have the inside bar of shoe straight and longer than the outside; also have the inside heel caulk run lengthwise instead of sidewise, as it usually does, and it should be at least $\frac{1}{4}$ inch longer than the outside one, and in some cases $\frac{3}{8}$ inch longer.

Horseshoe Vise.—Fig. 9 shows a home-made vise for holding horseshoes while sharpening, or shaping, which is a very useful tool. It is quite a task to sharpen a toe caulk on the anvil. It slips off and the

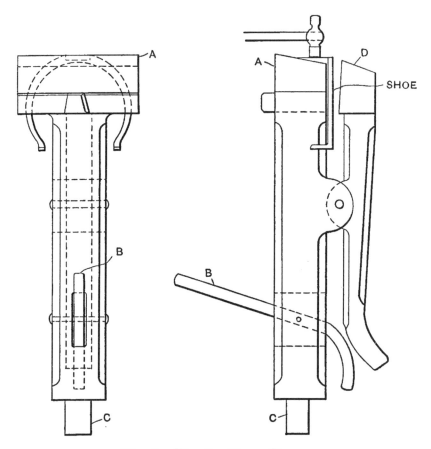

Fig. 9. Vise for Horse Shoes.

smith knocks a chunk off the anvil edge. Anvils used in this way look like a piece of cast-iron scrap fastened to a block. This vise can be made from cast iron, all but the tool steel jaw *A* and the machine steel lever *B*. The vise is held by spiking a piece of hard wood plank to the floor and fastening a plate of machine steel with a square hole in its center on top of the plank, into which the square shank C fits. After welding the toe caulk on a shoe, place shoe between jaws of vise and put foot on lever B and press down. This will bring the jaw against the shoe. Then draw the toe to the desired thinness. Jaw A can be so shaped as to give the caulk the required forward incline.

Shoeing for Contracted Feet.—Contracted feet are very painful to the horse and a source of annoyance

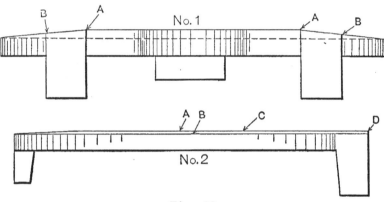

Fig. 10.

to the owner, but the shoer can do a great deal to lessen this trouble. Do not pare the hoof too much, and the shoe should not be fitted to the foot by burning. This is a horrible practice among shoers. In shaping the shoe get it level on top. Have the bar higher at the inside than at outside. The bar of shoe should taper about $\frac{1}{8}$ inch from inside to outside, A to B, Fig. 10, which gives a side and back view of the shoe. This shoe should be made from tool steel and hardened on top of bar from C to D, or about one third of length of bar. This prevents the hoof from growing into the shoe, as it very often does. It is not a good idea to remove a shoe too often; it should be left on for at least two months. By shaping the shoe, and hardening, as just described, the hoof will tend to spread out wider and the trouble is soon overcome.

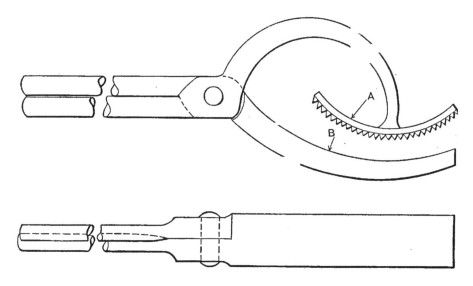

Fig. 11. Clinching Pinchers.

A foot which is contracted is very tender, as every horseman knows, and the horse can stand very little hammering or pounding on the hoof when clinching the nails. Some shoers are cruel enough to pound the animal in the ribs with the hammer for flinching under the painful operation of hammering the nails to place in the old fashioned way. Fig. 11 shows a pair of clinching pinchers that any smith can make for this purpose. They can be made from soft machine steel, except the part with teeth A. The pinchers are shown closed in the sketch. By opening and placing jaw B under the nail head and closing down on the pinchers, the teeth *A* on the jaw will do the work nicely and scarcely any hammering is necessary, thus saving the shoer any unnecessary trouble and the horse a deal of suffering.

ADVICE TO FOREMEN.

When hiring a man, tell him what you expect of him. If he does not come up to the standard, after repeated trials, let him go. Do not abuse him.

Have as little as possible to say to your men, and that on business only.

Do not take a piece of work to a smith and tell him how it should be done. Perhaps he knows more about it than you do; but if you see that he does not understand it, then a little advice is all right.

Keep a tidy shop by being tidy yourself.

Don't sit down yourself and issue orders for the men to stand.

If you promise a man an increase of wages, keep your promise, for if he finds you unreliable he will have no respect for you.

If you find a man is worthy, raise his wages. Don't wait for him to ask for an increase, as some men would rather quit than ask for an increase of wages.

Don't come in late yourself and abuse your men for being late once in a while.

Don't abuse a man for not doing enough work if you see he is killing time; let him go and put another in his place.

A man who is always watching his foreman is a good man to let go.

Don't hire a man who smokes cigarettes, as there are plenty of good men who do not smoke them.

If conditions at home or up town are not pleasant, don't lay out your spite on the men—they are not to blame.

Don't lay off a man who has a family to support and keep one who is inferior and has no one depending on him, and who perhaps wastes his money foolishly.

A foremanship earned by hard work and strict attention to business lasts longer and is better than one stolen by underhanded methods. I have worked my way up to every good position I have held.

J. F. SALLOWS.

APPENDIX

Decimal Equivalents.

4ths, 8ths, 16ths, 32nds, and 64ths of an inch.

Decimal	16	32	64	Decimal	16	32	64
.015625			1	.515625			33
.03125		1		.53125		17	
.046875			3	.546875			35
.0625	1			.5625	9		
.078125			5	.578125			37
.09375		3		.59375		19	
.109375			7	.609375			39
.125	2		.	.625	10		
.140625			9	.640625			41
.15625		5		.65625		21	
.171875			11	.671875			43
.1875	3			.6875	11		
.203125			13	.703125			45
.21875		7		.71875		23	
.234375			15	.734375			47
.25	4			.75	12		
.265625			17	.765625			49
.28125		9		.78125		25	
.296875			19	.796875			51
.3125	5			.8125	13		
.328125			21	.828125			53
.34375		11		.84375		27	
.359375			23	.859375			55
.375	6			.875	14		
.390625			25	.890625			57
.40625		13		.90625		29	
.421875			27	.921875			59
.4375	7			.9375	15		
.453125			29	.953125			61
.46875		15		.96875		31	
.484375			31	.984375			63
.5	8						

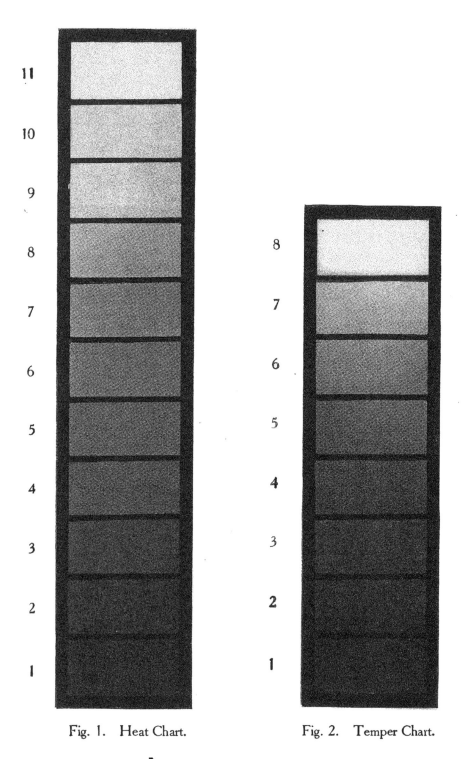

Fig. 1. Heat Chart.   Fig. 2. Temper Chart.

See Chapter III for Explanation of Charts and Methods of Using Them.

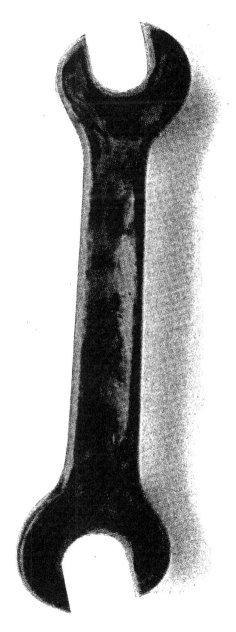

Fig. 3. Reproduction of Wrench showing Effects obtained by Methods for Casehardening and Coloring described in Chapter V.

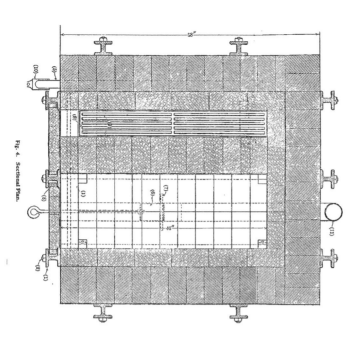

Fig. 4. Sectional Plan.

WORKING DRAWINGS
OF
COAL BURNING
CASEHARDENING FURNACE
(For Names of Numbered Parts see Chapter V)

INDEX

Advice to foremen..........147
Air hardening, device for.....110
Angle, making a double...... 28
Annealing tool steel..........113
    high-speed steel..........117
Anvil block.................. 4
Bath for hardening tools..... 83
Bending forks................ 13
Bending gas pipe............140
Bevel set, use of 14, 44
Boring tool.................. 53
    to lengthen............... 55
Box for packing pieces for casehardening..........120
    for pack hardening.......126
Brass turning tool........... 62
Brazing 134
    burner for..............136
    cast iron................139
    directions for...........137
    furnace 134
    small band saw..........139
    spelter for..............136
Broaches, hardening.......... 95
Burner for brazing..........136
Bushings, hardening..........101
Butt weld.................... 33
Cape chisels................. 46
Casehardening 118
    directions for...........121
    furnace 119
    packing for..............120
    tank for.................122
Centering tools.............. 58
Charcoal fire................ 73
Charred bone and leather, to obtain 132
Chisels:
    cape 46
    cold 44
    diamond point........... 48
    grooving 46
    hardening 79

Chisels:
    round-nose 47
Cleft weld................... 33
Cold chisels................. 46
    hardening 79
Color charts.............76, 152
Coloring steel...............131
    with cyanide............132
    by heat alone...........133
Crankshaft, making a........ 26
Cutters, see milling cutters.
Cutting off steel.............7-9
    high-speed steel..........108
    cutting-off tool, making.. 49
Cyanide, hardening with..112-130
    coloring with............132
Diamond point chisel......... 48
Diamond point tools......... 59
Dies, hardening............92-94
    to make without welding.. 63
Dipping tools in hardening...
    80-82-86
Drip pan.................... 87
Drawing temper, see tempering.
Dutchmen, use of............ 39
    Finishing tools.......... 58
Flux for welding............. 29
Forges, arrangement of....... 5
    see furnaces.
Forging 8
    bending fork............ 13
    blacksmith's tongs....... 10
    cape chisels............. 46
    cold chisels............. 44
    crankshaft 26
    double angle............ 28
    grooving chisels, etc...... 46
    heading tool............. 13
    high-speed steel..........109
    key-puller 17
    lathe and planer tools:
        boring tools........... 52

INDEX

Forging:
  lathe and planer tools:
    brass turning tool...... 62
    centering tools......... 58
    cutting-off tools........ 49
    diamond-point tool..... 59
    finishing tools......... 58
    roughing tools......... 56
    round-nose tool........ 62
    side tool.............. 51
    threading tools......... 55
  planer bolts............. 16
  rings or dies without welding 63
  rock drills.............. 62
  self-hardening steel.......109
  socket wrench........... 21
  spanner wrench.......... 23
  square corner on heavy stock 26
  turnbuckle or swivel..... 23
  wrenches 19
Furnaces:
  brazing 134
  casehardening 118
  for high-speed steel......113
  for tools................ 74
Grooving chisels............. 46
Hammer head, hardening.....102
Hardening 66
  broaches 95
  center punches.......... 81
  cold chisels............. 79
  drill jig bushings........101
  formed cutters.......... 85
  general directions for....105
  hammer head...........102
  high-speed steel..........111
  large rolls.............. 99
  lathe and planer tools.... 81
  milling cutters........... 83
    high-speed steel........112
  punches and dies....81-92-94
  reamers 88
  shear blades............ 97
  taps 91
  thin cutters...........87-128
  with cyanide.........112-130

Heading tool................ 13
Heating:
  care in................. 43
  high-speed steel..........111
  method of............... 8
  tool steel............... 67
Heat chart..................152
High-speed steel.............107
  annealing 117
  cutting off.............107
  forging 109
  hardening 110
  long blades............115
  milling cutters........112
  with cyanide..........112
  to distinguish...........107
Horseshoeing 142-143-145
Horseshoe vise..............144
Jimmie-bar 17
Key puller.................. 17
Lathe tools, see tools.
Lap weld................... 33
Machine forging............. 1
Mottling, see coloring.
Micro-photographs 65
Milling cutters:
  hardening 83-85-87-128
  high-speed steel........112
Oven, home made............ 74
Open-end wrench............ 19
Pack hardening..............125
  boxes for..............126
  long pieces............. 89
  testing work............127
  thin cutters............128
  reamers 89
Patterns for tools........... 43
Planer bolts................ 16
Planer tools, see tools.
Plow shares, repairing........142
Pinchers for clinching horse nails 146
Punches, hardening.........81-92
Pipe, welding ends in........137
  bending 140
Reading drawings............ 1
Reamers, hardening.......... 88
Ring, welding a............. 34

INDEX

Ring:
    to make without welding.. 63
Rock drill.................... 62
Rolls, hardening............. 99
Round-nose tool.............. 62
Roughing tool................ 56
Round-nose chisel............ 46
Scarf weld................... 30
Screw driver................. 48
Self-hardening steel, treatment
    of 109
Shear-blades, hardening....... 97
    high-speed steel..........115
Sheet steel, straightening.....141
Shoeing to prevent interfering 143
    for contracted feet.......145
Shrinkage and expansion.....102
Side tools................... 51
Spanner wrench............. 23
Spelter 136
Springs, tempering...........100
Socket wrench............... 31
Straightening sheet steel......141
Swivel, directions for making.. 23
System for handling tool
    work 40
Taps, hardening............. 91
Tempering 70-78
    see hardening.
    cold chisels.............. 80
    fine steel points.........102

Tempering:
    punches and dies........ 94
    springs 100
Temper chart.............76-152
Threading tools.............. 55
Tools:
    boring 52
    brass turning........... 62
    centering 58
    finishing 58
    from old files........... 48
    of machine steel.........124
    hardening 79-81
    roughing 56
    round-nose 62
    side 51
    threading 55
Turnbuckles 23
Vise for horseshoers.........144
Welding:
    butt weld............... 33
    cleft weld............... 33
    directions for........... 29
    flux for................. 29
    jump weld.............. 33
    lap weld................ 33
    ring 34
    scarf weld.............. 30
    solid ends in pipe........ 37
    to avoid two heats for.... 31
    use of "dutchmen" in.... 39
Working drawings........... 1